INORGANIC SYNTHESES

Volume 23

Editor-in-Chief

STANLEY KIRSCHNER

Department of Chemistry
Wayne State University
Detroit, Michigan

INORGANIC SYNTHESES

Volume 23

A Wiley-Interscience Publication
JOHN WILEY & SONS

New York Chichester Brisbane Toronto Singapore

Published by John Wiley & Sons, Inc.

Library of Congress Catalog Number: 39-23015

ISBN 0-471-81873-9

Printed in the United States of America

10 9 8 7 6 5 4 3 2 1

MEMBERS

This volume is dedicated to the memory of

EARL L. MUETTERTIES
(1927–1984)

He was a chemist's chemist and a friend to
Inorganic Syntheses

PREFACE

I welcome this opportunity to contribute to *Inorganic Syntheses* by editing this volume, because this series has been of considerable help to me and my colleagues for more than thirty years. It is appropriate, I believe, to express my appreciation and the appreciation of *Inorganic Syntheses* to all those who were involved in producing this volume. Special thanks must go to the contributors and checkers, who provided and checked the substantive material of the volume. However, additional thanks must go to the members of *Inorganic Syntheses* who contributed their time and expertise to help with the editing, because, without this assistance, this volume would not be nearly so complete as it is.

I would like especially to thank John Bailar, Therald Moeller, and Thomas Sloan, who were particularly helpful during the editing process. Further, I would also like to express my appreciation to Duward Shriver, for his help in providing guidance and for his editorial assistance, as well. I must extend my sincere appreciation to my secretary, Elena Drob, whose patience and efforts beyond the call of duty helped to make this volume possible.

I wish to call special attention to the **Special Hazard Notice** Appendix at the back of this volume.

Finally, I wish to mention the dedication of this volume to Earl Muetterties, whose untimely passing has saddened us all. He was the Editor of Volume 10 of this series and continued to contribute to the success of later volumes.

STANLEY KIRSCHNER

Detroit, Michigan
December 1984

Previous volumes of *Inorganic Syntheses* are available. Volumes I–XVI can be ordered from R. E. Krieger Publishing Co., Inc., P.O. Box 9542, Melbourne, Florida 32901; Volume XVII is available from McGraw-Hill, Inc.; and Volumes XVIII–XXII can be obtained from John Wiley & Sons, Inc.

NOTICE TO CONTRIBUTORS AND CHECKERS

The *Inorganic Syntheses* series is published to provide all users of inorganic substances with detailed and foolproof procedures for the preparation of important and timely compounds. Thus, the series is the concern of the entire scientific community. The Editorial Board hopes that all chemists will share in the responsibility of producing *Inorganic Syntheses* by offering their advice and assistance in both the formulation of and the laboratory evaluation of outstanding syntheses. Help of this kind will be invaluable in achieving excellence and pertinence to current scientific interests.

There is no rigid definition of what constitutes a suitable synthesis. The major criterion by which syntheses are judged is the potential value to the scientific community. An ideal synthesis is one that presents a new or revised experimental procedure applicable to a variety of related compounds, at least one of which is critically important in current research. However, syntheses of individual compounds that are of interest or importance are also acceptable. Syntheses of compounds that are readily available commercially at reasonable prices are not acceptable. Corrections and improvements of syntheses already appearing in *Inorganic Syntheses* are suitable for inclusion.

The Editorial Board lists the following criteria of content for submitted manuscripts. Style should conform with that of previous volumes of *Inorganic Syntheses*. The introductory section should include a concise and critical summary of the available procedures for synthesis of the product in question. It should also include an estimate of the time required for the synthesis, an indication of the importance and utility of the product, and an admonition if any potential hazards are associated with the procedure. The Procedure should present detailed and unambiguous laboratory directions and be written so that it anticipates possible mistakes and misunderstandings on the part of the person who attempts to duplicate the procedure. Any unusual equipment or procedure should be clearly described. Line drawings should be included when they can be helpful. All safety measures should be stated clearly. Sources of unusual starting materials must be given, and, if possible, minimal standards of purity of reagents and solvents should be stated. The scale should be reasonable for normal laboratory operation, and any problems involved in scaling the procedure either up or down should be discussed. The criteria for judging the purity of the final product should be delineated clearly. The section on Properties should supply and discuss those physical and chemical characteristics that are relevant to judging the purity of the product and to permitting its handling and use in an intelligent manner. Under References, all pertinent literature citations should be listed in order. A style sheet is available from the Secretary of the Editorial Board.

The Editorial Board determines whether submitted syntheses meet the general specifications outlined above. Every procedure will be checked in an independent laboratory, and publication is contingent upon satisfactory duplication of the syntheses.

Each manuscript should be submitted in duplicate to the Secretary of the Editorial Board, Professor Jay H. Worrell, Department of Chemistry, University of South Florida, Tampa, FL 33620. The manuscript should be typewritten in English. Nomenclature should be consistent and should follow the recommendations presented in *Nomenclature of Inorganic Chemistry,* 2nd Ed., Butterworths & Co., London, 1970 and in *Pure Appl. Chem.,* **28,** No. 1 (1971). Abbreviations should conform to those used in publications of the American Chemical Society, particularly *Inorganic Chemistry.*

Chemists willing to check syntheses should contact the editor of a future volume or make this information known to Professor Worrell.

TOXIC SUBSTANCES AND LABORATORY HAZARDS

Chemicals and chemistry are by their very nature hazardous. Chemical reactivity implies that reagents have the ability to combine. This process can be sufficiently vigorous as to cause flame, an explosion, or, often less immediately obvious, a toxic reaction.

The obvious hazards in the syntheses reported in this volume are delineated, where appropriate, in the experimental procedure. It is impossible, however, to foresee every eventuality, such as a new biological effect of a common laboratory reagent. As a consequence, *all* chemicals used and *all* reactions described in this volume should be viewed as potentially hazardous. Care should be taken to avoid inhalation or other physical contact with all reagents and solvents used in this volume. In addition, particular attention should be paid to avoiding sparks, open flames, or other potential sources which could set fire to combustible vapors or gases.

A list of 400 toxic substances may be found in the *Federal Register,* Vol. 40, No. 23072, May 28, 1975. An abbreviated list may be obtained from *Inorganic Syntheses,* Vol. 18, p. xv, 1978. A current assessment of the hazards associated with a particular chemical is available in the most recent edition of *Threshold Limit Values for Chemical Substances and Physical Agents in the Workroom Environment* published by the American Conference of Governmental Industrial Hygienists.

The drying of impure ethers can produce a violent explosion. Further information about this hazard may be found in *Inorganic Syntheses*, Vol. 12. p. 317.

See **Special Hazard Notice** Appendix at the back of this volume.

CONTENTS

Chapter One ORGANOMETALLIC COMPOUNDS

Chapter Two COMPOUNDS OF BIOLOGICAL INTEREST

Chapter Three STEREOISOMERS

Chapter Four BRIDGE AND CLUSTER COMPOUNDS

Chapter Five UNUSUAL LIGANDS AND COMPOUNDS

INORGANIC SYNTHESES

Volume 23

Chapter One

ORGANOMETALLIC COMPOUNDS

1. (SUBSTITUTED THIOUREA)PENTACARBONYLCHROMIUM(0) COMPLEXES

Submitted by J. A. COSTAMAGNA* and J. GRANIFO†
Checked by MARCETTA Y. DARENSBOURG‡ and PATRICIA A. TOOLEY‡

There are several conventional methods for synthesizing mixed ligand complexes of the type $[Cr(CO)_{6-x}L_x]$, where L is a phosphine or arsine-type ligand.[1] In these cases, when S-donor ligands are used, sulfur-bridged compounds are generally formed. It has been possible to isolate compounds of formula $[Cr(CO)_5L]$ where L is a two-electron-pair S-donor ligand.[2,3] The method consists of the irradiation of solutions of metal carbonyls in tetrahydrofuran (THF) followed by addition of the ligand.[4] It is possible to apply the same procedure to molybdenum and tungsten carbonyls with various substituted thioureas.

Procedure

$$Cr(CO)_6 + THF \xrightarrow{h\nu} [Cr(CO)_5(THF)] + CO$$

All the reactions are performed under an atmosphere of dry nitrogen, free of oxygen; Pyrex glass and standard ground joints are used. The equipment in

*Departamento de Quimica, Universidad Tecnica del Estado, Santiago, Chile.
†Facultad de Ciencias, Universidad de Chile, Santiago, Chile.
‡Department of Chemistry, Texas A & M University, College Station, TX 77843.

which this reaction is carried out consists of a Pyrex glass cylinder with three exits, two lateral and one superior. Inside the cylinder there is an ultraviolet immersion lamp (original Hanau, medium pressure), protected by a quartz double wall.

Hexacarbonylchromium(0) (660.0 mg, 3 mmoles) is placed in the reaction vessel* and dissolved in 60 mL of dry peroxide-free tetrahydrofuran (THF);[5–7] this is irradiated for one hour and an orange solution is obtained, which is placed (avoid air contact) in a 100-mL round-bottomed flask with standard connections for work under an inert atmosphere.

An equal volume of this solution is used as the starting material for all of the syntheses. The yield is 30–40%.

A. PENTACARBONYL(THIOUREA)CHROMIUM(0)

$$[Cr(CO)_5THF] + (NH_2)_2CS \longrightarrow [Cr(CO_5\{(NH_2)_2CS\}] + THF$$

To the orange solution of $[Cr(CO)_5(THF)]$ 228.4 mg of thiourea* (3 moles) are added and the system is stirred magnetically for 10 minutes to complete the reaction. The solution changes to a yellowish color. Tetrahydrofuran is eliminated by evaporation at room temperature and reduced pressure; at the same time, under these conditions, the excess hexacarbonylchromium sublimes in approximately one hour and is collected on a cold finger. The resulting crude product is dissolved in 50 mL of benzene and filtered through a 5-cm Kieselguhr column supported by a glass filtration frit. The yellow filtrate is received in a 150 mL flask. The volume is then reduced under vacuum until a yellow precipitate starts to appear. An abundant crop of crystals is obtained after the addition of petroleum ether (60–80° fraction). These are collected by filtration on a frit and are washed twice with petroleum ether (60–80° fraction). The crystals are vacuum-dried at room temperature. The yellow crystals are needle shaped. *Anal.* Calcd. for $C_6H_4N_2O_5SCr$: C. 26.86; H, 1.50; N, 10.45. Found: C, 26.98; H, 1.51; N, 10.63.

B. PENTACARBONYL(N,N,N′,N′-TETRAMETHYLTHIOUREA)CHROMIUM(0)

$$[Cr(CO)_5(THF)] + (CH_3)_4N_2CS \longrightarrow [Cr(CO)_5\{(CH_3)_4N_2CS\}] + THF$$

To the orange solution of $[Cr(CO)_5(THF)]$, 317.4 mg of tetramethyl thiourea* (2.4 mmoles) is added and the resulting solution is stirred for 10 minutes. The

*Benzene and thiourea and its methyl-substituted analogs have been identified as carcinogens and must be handled with gloves and with care. Also, metal carbonyls are toxic and must be handled with care in a hood.

excess THF and hexacarbonylchromium are eliminated as above. The crude solid is dissolved in 15 mL of a 2:1 volume mixture of benzene:heptane and the solution is filtered through a Kieselguhr column. The resulting yellowish solution is concentrated to 2 mL. Then 5 mL of petroleum ether, 40–60° fraction, are added and the precipitate is separated by decantation. The yellow-orange crystals are filtered and treated as before. *Anal.* Calcd. for $C_{10}H_{12}N_2O_5SCr$: C, 37.01; H, 3.73; N, 8.64. Found: C, 37.34; H, 3.89; N, 8.63.

C. PENTACARBONYL(N,N′-DI-*P*-TOLYLTHIOUREA)CHROMIUM(0)

$$[Cr(CO)_5(THF)] + (p\text{-}CH_3C_6H_4NH)_2CS \longrightarrow [Cr(CO)_5\{(p\text{-}CH_3C_6H_4NH)_2CS\}] + THF$$

To the solution of $[Cr(CO)_5(THF)]$ 609.3 mg of N,N′-di-*p*-tolylthiourea (2.4 mmoles) is added and the resulting solution is treated as in the previous cases. The crude product is dissolved in 75 mL of benzene and is filtered through a Kieselguhr column. The solution is vacuum-concentrated in vacuo to 10 mL; then 20 mL of petroleum ether, 40–60° fraction, is added and the yellow precipitate is separated by decantation, filtered, washed, and dried as before. *Anal.* Calcd. for $C_{20}H_{16}N_2O_5SCr$; C, 53.55; H, 3.60; N, 6.25. Found: C, 53.61; H, 3.57, N, 6.17.

D. PENTACARBONYL(N,N′-DI-*TERT*-BUTYLTHIOUREA)CHROMIUM(0)

$$[Cr(CO)_5(THF)] + (t\text{-}C_4H_9NH)_2CS \longrightarrow [Cr(CO)_5\{(t\text{-}C_4H_9NH)_2CS\}] + THF$$

To the solution of $[Cr(CO)_5(THF)]$ 452 mg of N,N′-di-*tert*-butylthiourea is added and the resulting solution is treated as before. The crude product is dissolved in 40 mL of benzene and then 40 mL of petroleum ether, 40–60° fraction, is added. The solution is filtered through a Kieselguhr column, and the solvent is removed completely under vacuum. Then 5 mL of petroleum ether, 40–60° fraction, is added and the suspension is filtered through a frit. The yellow crystals are treated as before. *Anal.* Calcd. for $C_{14}H_{20}N_2O_5SCr$: C, 44.18; H, 5.30; N, 7.37. Found: C, 44.32; H, 5.39; N, 7.45.

Properties

The complexes are yellow crystalline solids (needles) and they decompose before melting. They are moderately stable in air but relatively unstable in benzene, chloroform, and diethyl ether solutions. They are insoluble in petroleum ether. The molecular weights of the complexes, obtained by cryoscopy in benzene under a nitrogen atmosphere, are consistent with a monomeric character in solution. Similar techniques for the preparation of other group VI metal carbonyls have been reported by Tripathi and co-workers[8] and Lindner and Nagel.[9]

References

1. E. W. Abel and F. G. A. Stone, *Quart. Rev.*, **24,** 498 (1970).
2. E. W. Aiscough, E. J. Bird, and A. M. Brodie, *Inorg. Chim. Acta,* **20,** 187 (1976).
3. E. W. Aiscough, A. M. Brodie, and A. R. Furness, *J. Chem. Soc.* (*Dalton*) 2360 (1972).
4. W. Strohmeier, *Agnew. Chem. Intern. Ed.* (*Engl.*), **3,** 730 (1964).
5. R. W. Parry (ed.), *Inorganic Syntheses,* Vol. 12, McGraw-Hill Book Co., New York, 1970, p. 317.
6. J. Granifo, J. Costamagna, A. Garrao, and M. Pieber, *J. Inorg. Nucl. Chem.*, **42,** 1587 (1980).
7. F. A. Cotton and C. S. Kraihanzel, *J. Am. Chem. Soc.*, **84,** 4432 (1962).
8. S. C. Tripathi, S. C. Srivastava, and R. D. Pandey, *J. Inorg. Nucl. Chem.*, **35,** 457 (1973).
9. E. Lindner and W. Nagel, *Z. Naturforsch.*, **32b,** 1116 (1977).

2. DICARBONYLNITROSYL{TRIS(3,5-DIMETHYLPYRAZOLYL)HYDROBORATO}-MOLYBDENUM(III) AND IODO-, ALKOXY-, AND ALKYLAMIDO-MOLYBDENUM(III) DERIVATIVES

Submitted by S. J. REYNOLDS, C. F. SMITH, C. J. JONES, and J. A. McCLEVERTY*
Checked by D. C. BROWER and J. L. TEMPLETON†

Tris(3,5-dimethylpyrazolyl)hydroborate, $[HB(C_3H(CH_3)_2N_2)_3]^-$, is a uninegative, tridentate, six-electron, donor ligand formally isoelectronic with η^5-$C_5H_5^-$ in its bonding to metals. However, it is also a sterically bulky ligand, and it usually confers greater stability on its complexes than does η^5-$C_5H_5^-$. An interesting demonstration of these steric effects is provided by the molybdenum(III) complexes:[1] $Mo[HB\{C_3H(CH_3)_2N_2\}_3](NO)XY$, X = Cl, Br, I, OR; Y = Cl, Br, I, OR, NHR^1, and R = alkyl; R^1 = H, alkyl, aryl, NH_2, or NR_2. These complexes are unusual examples of stable, formally 16-electron, six-coordinate molybdenum(III) compounds. In contrast, the cyclopentadienyl analogs[2,3] are isolated as dimers or ligand adducts which are 18-electron compounds of seven-

*Department of Chemistry, University of Birmingham, Birmingham B15 2TT, United Kingdom.
†Department of Chemistry, University of North Carolina at Chapel Hill, Chapel Hill, NC 27514.

coordinate molybdenum(III) (counting η^5-C_5H_5 as occupying three coordination sites).

The $Mo[HB\{C_3H(CH_3)_2N_2\}_3](NO)(CO)_2$ is a convenient starting material for preparing these stable, 16-electron compounds, and it may be obtained from $Mo(CO)_6$ in good yield by the reaction sequence described below. It reacts with iodine to give $Mo[HB\{C_3H(CH_3)_2N_2\}_3](NO)I_2$, the preparation of which has been reported earlier, but an improved synthesis is now described.[1] The preparation of the ethoxy-[1] and ethylamido-[4] derivatives of this diiodide are also described to provide examples of its substitution reactions. Related derivatives may be prepared by similar methods.

A. DICARBONYLNITROSYL{TRIS(3,5-DIMETHYLPYRAZOLYL)BORATO}MOLYBDENUM(III)

$$[Mo(CO)_6] + K[HB\{C_3H(CH_3)_2N_2\}_3] \longrightarrow K[Mo[HB\{C_3H(CH_3)_2N_2\}_3](CO)_3] + 3CO\uparrow$$

$$K[Mo[HB\{C_3H(CH_3)_2N_2\}_3](CO)_3] + CH_3CO_2H \longrightarrow Mo[HB\{C_3H(CH_3)_2N_2\}_3](CO)_3H + CH_3CO_2K$$

$$Mo[HB\{C_3H(CH_3)_2N_2\}_3](CO)_3H + CH_3C_6H_4SO_2N(CH_3)NO \longrightarrow Mo[HB\{C_3H(CH_3)_2N_2\}_3](CO)_2NO + CH_3C_6H_4SO_2NHCH_3 + CO\uparrow$$

Procedure

■ **Caution.** *Although the final reaction product is air stable, the intermediates are very oxygen-sensitive. This reaction must be performed under nitrogen using nitrogen-saturated tetrahydrofuran (THF). Prior to use, THF is distilled from sodium/benzophenone under nitrogen. Safe procedures for purifying THF have been described previously:* Inorganic Syntheses, **12,** *317* (*1970*).*

A solution of $K[HB\{C_3H(CH_3)_2N_2\}_3]$ (15.5 g)[5] and $[Mo(CO)_6]$ (12 g) in freshly distilled THF (200 mL) is prepared in a 500-mL nitrogen-filled, conical flask fitted with a reflux condenser and magnetic stirrer. The mixture is refluxed overnight (~18 hours) under nitrogen and stirred on a stirrer hotplate. The resulting yellow suspension is then allowed to cool before adding glacial acetic acid (5 mL) and stirring for 1.5 hours more.† A solution of N-methyl-N-nitroso-

*The yields in this reaction appear to be somewhat dependent on the quality of THF used. Low yields can be obtained using one batch of THF despite careful purification. The only way to overcome this is to use a different batch of solvent.

†The reaction also works if the addition of acetic acid is omitted. However, a reduced yield (~64%) is obtained.

p-toluene sulfonamide* (9.8 g) in freshly distilled THF (50 mL) is then added and the mixture is stirred at room temperature overnight (~20 hours). (The subsequent procedures for extracting the product from the resulting orange suspension may be carried out in air.) The solvent is then evaporated under reduced pressure (using a rotary evaporator) to give an orange solid. This solid is redissolved in chloroform (~200 mL) and the solution is filtered through Kieselguhr to remove undissolved toluene sulfonamide. All the chloroform is then removed using a rotary evaporator, and the crude, solid product is washed with ethanol (~three × 100 mL), recrystallized from chloroform/ethanol, and dried under vacuum. Yield 15–17 g (71–78%). *Anal.* Calcd. for $MoC_{17}H_{22}N_7O_3B$: C, 42.62; H, 4.60; N, 20.47. Found: C, 42.20; H, 4.85; N, 20.00.

Properties

Dicarbonyl-nitrosyl{tris(3,5-dimethylpyrazolyl)hydroborato}molybdenum(III) is an orange, crystalline solid, soluble in chloroform and dichloromethane, but insoluble in ethanol, methanol, and hexane. It is air stable, but is best stored for prolonged periods under dry nitrogen. The infared spectrum (KBr disc) shows the following characteristic bands: ν_{NO} 1655 cm^{-1}; ν_{CO} 1906, 2006 cm^{-1}; ν_{BH} 2546 cm^{-1}. The 220 MHz ^1H nmr spectrum in $CDCl_3$ shows two signals at δ_{TMS} = 5.81 (area 2) and 5.72 (area 1), attributable to the pyrazolyl protons, along with signals at δ_{TMS} = 2.46 (area 6), 2.35, and 2.32 (area 12), attributable to the methyl protons.

B. DIIODONITROSYL{TRIS(3,5-DIMETHYLPYRAZOLYL)HYDROBORATO}MOLYBDENUM(III)

$$Mo[HB\{C_3H(CH_3)_2N_2\}_3](NO)(CO)_2 + I_2 \longrightarrow Mo[HB\{C_3H(CH_3)_2N_2\}_3](NO)I_2 + 2CO\uparrow$$

Procedure

Iodine (6.0 g) is placed in a Soxhlet thimble and extracted into a refluxing solution of $Mo[HB\{C_3H(CH_3)_2N_2\}_3](NO)(CO)_2$ (12.0 g) in methylcyclohexane (~250 mL). After 72 hours the mixture is allowed to cool, and the product is isolated by filtration. It is then recrystallized from boiling toluene to give black crystals of the mono-toluene solvate. Yield: 11.5–12.5 g (60–65%). *Anal.* Calcd.

*Obtained from Fisons Ltd., U.K.

for $MoC_{22}H_{30}N_7OBI_2$: C, 34.34; H, 3.90; N, 12.75. Found: C, 34.53; H, 3.92; N, 12.94.

Properties

Diiodo{tris(3,5-dimethylpyrazolyl)hydroborato}molybdenum(III) is an air-stable, black, crystalline solid that dissolves in chloroform, dichloromethane, and toluene, but is only slightly soluble in hexane. It reacts with alcohols (see below), and is best stored for prolonged periods under dry nitrogen. The infrared spectrum (KBr disc) shows ν_{NO} 1700 cm^{-1} and ν_{BH} 2555 cm^{-1}, and the 220 MHz ^1H nmr spectrum in $CDCl_3$ shows signals attributable to the pyrazolyl protons at δ_{TMS} = 6.09 (area 1) and 5.86 (area 2), the phenyl protons at δ_{TMS} = 7.13, 7.27 (area 5), and the pyrazolyl methyl protons at δ_{TMS} = 2.77 (area 3), 2.58 (area 3), 2.29 (area 6), and 2.19 (area 6).

C. ETHOXYIODONITROSYL{TRIS(3,5-DIMETHYLPYRAZOLYL)HYDROBORATO}MOLYBDENUM(III)

$$Mo[HB\{C_3H(CH_3)_2N_2\}_3](NO)I_2 + C_2H_5OH \longrightarrow Mo[HB\{C_3H(CH_3)_2N_2\}_3](NO)I(OC_2H_5) + HI$$

Procedure

A sample of $Mo[HB\{C_3H(CH_3)_2N_2\}_3](NO)I_2 \cdot C_6H_5CH_3$ (1.0 g) is refluxed in ethanol (40 mL) for one hour, during which time the solution turns dark green. The mixture is allowed to cool and the crude product is collected by filtration. It may be recrystallized from dichloromethane/ethanol. Yield: 0.7 g (90%). *Anal.* Calcd. for $MoC_{17}H_{27}N_7BO_2I$: C, 34.30; H, 4.54; N, 16.48. Found: C, 34.6; H, 4.7; N, 16.4.

Properties

Ethoxy-iodo-nitrosyl{tris(3,5-dimethylpyrazolyl)hydroborato}molybdenum(III) is an air-stable, green, crystalline solid, soluble in chloroform and dichloromethane. It is insoluble in hexane and slightly soluble in cold alcohols, with which it undergoes exchange reactions. The infrared spectrum (KBr disc) exhibits ν_{NO} 1678 cm^{-1} and ν_{BH} 2552 cm^{-1}. The 220 MHz ^1H nmr spectrum in $CDCl_3$ contains signals attributable to pyrazolyl protons at δ_{TMS} = 5.86, 5.83, and 5.79 (all area 1), and pyrazolyl methyl protons at δ_{TMS} = 2.51, 2.43, 2.39, and 2.35

(overall area 18). Signals due to the ethyl group appear at δ_{TMS} = 5.68 (area 2, AB pair) and 1.57 (area 3, triplet).

D. ETHYLAMIDOIODONITROSYL{TRIS(3,5-DIMETHYLPYRAZOLYL)HYDROBORATO}MOLYBDENUM(III)

$$Mo[HB\{C_3H(CH_3)_2N_2\}_3](NO)I_2 + 2C_2H_5NH_2 \longrightarrow Mo[HB\{C_3H(CH_3)_2N_2\}_3](NO)I(NHC_2H_5) + C_2H_5NH_3I$$

Procedure

■ **Caution.** *Diisopropyl ether forms a solid, explosive peroxide on prolonged contact with the air. Only fresh material known to be peroxide-free should be used.*

A solution of $Mo[HB\{C_3H(CH_3)_2N_2\}_3](NO)I_2 \cdot C_6H_5CH_3$ (0.5 g) and an excess of ethylamine (0.1 mL) in dichloromethane (40 mL) is stirred at room temperature for 0.5 hour during which time the solution becomes deep red. The volume of the solution is then reduced by evaporation (to ~20 mL), and diisopropyl ether is added dropwise to produce a white precipitate of ethylammonium iodide. The mixture is then filtered and the dichloromethane is removed from the filtrate by evaporation, causing the crude, red product to precipitate from the diisopropyl ether. This is collected by filtration and recrystallized from dichloromethane/diisopropyl ether. Yield: 0.3 g (78%). *Anal.* Calcd. for $MoC_{17}H_{28}N_8BOI$: C, 34.36; H, 4.72; N, 18.86. Found: C, 34.61; H, 4.72; N, 18.61.

Analogous complexes derived from other primary amines or unsymmetrical hydrazines may be prepared by similar means, but, for steric reasons, secondary amines, other than $(CH_3)_2NH$, do not form stable complexes. The Mo-N bond can be cleaved by HCl to give $Mo[HB\{C_3H(CH_3)_2N_2\}_3](NO)ICl$. Mixed alkoxy-amido compounds of the form $Mo[HB\{C_3H(CH_3)_2N_2\}_3](NO)(OC_2H_5)(NHR)$ can be prepared in a similar manner if $Mo[HB\{C_3H(CH_3)_2N_2\}_3](NO)I(OC_2H_5)$ is substituted for the diiodide in the above reaction.

Properties

Ethylamido-iodo-nitrosyl{tris(3,5-dimethylpyrazolyl)hydroborato}molybdenum-(III) is an air-stable, red, crystalline solid, soluble in dichloromethane and chloroform. It reacts with alcohols to give alkoxy species with loss of the ethylamido ligand. The infrared spectrum (KBr disc) exhibits ν_{NO} 1660 cm^{-1}, ν_{BH} 2555 cm^{-1}, and ν_{NH} 3290 cm^{-1}. The 220 MHz ^{1}H nmr spectrum contains signals attributable to the pyrazolyl protons at δ_{TMS} = 5.90, 5.82, and 5.80 (all area 1), the pyrazolyl methyl protons at δ_{TMS} = 2.61, 2.59, 2.44, 2.36, and 2.33 (total area 18), and the amide ethyl group at δ_{TMS} = 4.50 (area 2, AB pair) and 1.40 (triplet, area 3). The amide NH proton signal is observed at δ_{TMS} = 12.49 (area 1, four lines), and is not collapsed by addition of D_2O.

References

1. J. A. McCleverty, D. Seddon, N. A. Bailey, and N. W. Walker, *J. Chem. Soc., Dalton Trans.*, 898 (1976).
2. T. A. James and J. A. McCleverty, *J. Chem. Soc.* (*A*), 3308 (1970). J. A. McCleverty and D. Seddon, *J. Chem. Soc., Dalton Trans.*, 2526 (1972).
3. D. Seddon, W. G. Kita, J. Bray, and J. A. McCleverty, *Inorg. Synth.*, **16,** 24 (1976).
4. J. A. McCleverty, A. E. Rae, J. Wolochowicz, N. A. Bailey, and J. M. A. Smith, *J. Chem. Soc., Dalton Trans.*, 429 (1982).
5. S. Trofimenko, *J. Am. Chem. Soc.*, **89,** 3170 and 6288 (1967).

3. DIISOCYANIDE COMPLEXES OF MOLYBDENUM(0) AND TUNGSTEN(0) AND DERIVED AMINOCARBYNE COMPLEXES

Submitted by A. J. L. POMBEIRO* and R. L. RICHARDS†
Checked by B. L. HAYMORE‡

Displacement of ligating dinitrogen by isocyanides constitutes a convenient route for the preparation of isocyanide complexes. The first isocyanide complexes prepared by this route[1] involve a molybdenum(0) or tungsten(0) electron-rich metal site, which activates the ligating isocyanide towards electrophilic attack. Aminocarbyne-type species are formed by reactions with acids[2,3] or alkylating agents.[4] The general methods for the preparation of the first isocyanide and derived terminal carbyne complexes obtained by this technique are given below. The nucleophilicity of the isocyanide ligand in those electron-rich complexes[5] is in contrast to its usual electrophilic character[6–8] reported for this ligand when it binds a metal ion in its normal or higher oxidation states.

*Centro de Química Estrutural, Complexo I, Instituto Superior Técnico, Lisboa 1000, Portugal.
†Unit of Nitrogen Fixation, The University of Sussex, Brighton, United Kingdom.
‡Corporate Research Laboratories, Monsanto Co., St. Louis, MO 63167.

A. *TRANS*-BIS[1,2-ETHANEDIYLBIS(DIPHENYLPHOSPHINE)]BIS(ISOCYANOMETHANE)TUNGSTEN(0)

$$trans\text{-}[W(N_2)_2(dppe)_2] + 2\ CNMe \longrightarrow trans\text{-}[W(CNMe)_2(dppe)_2] + 2\ N_2$$

$$(dppe = Ph_2PCH_2CH_2PPh_2)$$

Procedure

■ **Caution.** *Isocyanides (in particular, isocyanomethane) are extremely toxic and malodorous, and their manipulation should be performed in a well-ventilated hood with great care.*

The isocyanides can be prepared by the phosgene method of Ugi and coworkers,[9] except for CNMe[10] and $CNBu^t$,[11] and the dinitrogen *trans*-$[M(N_2)_2(dppe)_2]$ (M = Mo or W)[12] complexes can also be prepared by published methods.

The *trans*-$[W(N_2)_2(dppe)_2]$ (2.08 g, 2.0 mmoles) is dissolved in dry tetrahydrofuran (THF, 100 mL), under dinitrogen, in a 250-mL single-necked Schlenk flask. The solution is filtered under nitrogen to remove any insoluble impurities. Isocyanomethane (0.28 mL, 5.9 mmoles) is added under nitrogen to the filtered solution, which is then refluxed, with stirring, under nitrogen (or argon) for ~six hours. Concentration of the solution almost to dryness, under vacuum and with constant heating at ~50–60°, leads to the precipitation of *trans*-$[W(CNMe)_2(dppe)_2]$, as red prisms. The small volume of residual hot solution may be removed by decantation, and the red solid is then thoroughly washed, in air with several (5–10) portions (of ~10–15 mL each) of acetone, to remove a small amount of a yellow, powdery impurity. The pure, red crystals are then dried under vacuum. The yield is about 2.0 g (1.9 mmoles, 95%). *Anal.* Calcd. for $[W(CNMe)_2(dppe)_2]$: C, 63.3; H, 5.1; N, 2.6. Found: C, 63.5; H, 5.3; N, 2.7.

Properties

The product is a red, air-stable, crystalline solid. Its infared spectrum[1] has a strong, broad band at 1834 cm^{-1} (in a KBr disc), which is assigned to CN stretching. In the mass spectrum the parent ion is observed with a principal peak at m/e = 1062, as required for the predominant ^{184}W isotope of $[W(CNMe)_2(dppe)_2]^+$.

A series of analogous complexes of Mo(0) or W(0) with a variety of isocyanide ligands may be synthesized similarly:[1] *trans*-$[M(CNR)_2(dppe)_2]$ (M = Mo or W; R = Me, Bu^t, Ph, 4-MeC_6H_4, 4-$MeOC_6H_4$, 4-ClC_6H_4, 2,6-$Cl_2C_6H_3$. The dinitrogen ligand is more labile when ligating Mo(0) than W(0), and the quantitative formation of the diisocyanide complexes is generally faster for Mo(0) than for W(0) and, in the former, may even occur without heating if it is in the presence of sufficient excess of the appropriate isocyanide.

All the diisocyanide complexes are red (or deep red) in color, and they exhibit a strong, broad ν(CN) band in the 1788–1915 cm^{-1} region (in a KBr pellet).[1] Their electronic spectra have been recorded,[13] and their redox properties studied by cyclic voltammetry[1] and by chemical oxidation.[13] The molecular structure has been determined by X-ray diffraction[14] studies for the *trans*-[$Mo(CNMe)_2(dppe)_2$], with the isocyanides showing a CNC bond angle of 156(1)°. The cause of this bending is believed to be electronic in origin, and the electron-rich bent isocyanomethane in the Mo(O) and W(O) complexes is susceptible to attack by alkylating[4] or protonating[2,3] agents, the latter reactions being described below.

B. *TRANS*-BIS[1,2-ETHANEDIYLBIS(DIPHENYLPHOSPHINE)](ISOCYANOMETHANE)[(METHYLAMINO)METHYLIDYNE]TUNGSTEN(IV) TETRAFLUOROBORATE

$$trans\text{-}[W(CNMe)_2(dppe)_2] + [Et_2OH][BF_4] \longrightarrow trans\text{-}[W(CNHMe)(CNMe)(dppe)_2][BF_4] + Et_2O$$

Procedure

■ **Caution.** *Tetrafluoroboric acid is corrosive and toxic and concentrated solutions should be handled in a hood. Also see caution in Section 3-A.*

A solution of *trans*-[$W(CNMe)_2(dppe)_2$] (1.69 g, 1.59 mmoles) in benzene (180 mL) is prepared under nitrogen in a 250-mL single-necked Schlenk flask. The solution is filtered, under nitrogen, into a similar Schlenk flask, and tetrafluoroboric acid (1.50 mmoles; 4.40 mL in a diethyl ether solution prepared by 1:20 dilution of commercial [Et_2OH][BF_4]) is added dropwise, under nitrogen, to the vigorously stirred, filtered solution. The product precipitates as shiny, green needles, which are filtered under nitrogen, washed with benzene, and dried under vacuum. The yield is about 1.6 g (1.4 mmoles, 87% based on the parent tungsten complex). *Anal.* Calcd. for [$W(CNHMe)(CNMe)(dppe)_2$]BF_4: C, 58.5; H, 4.8; N, 2.4. Found: C, 58.5; H, 4.8; N, 2.4.

The use of an excess of acid must be avoided in this preparation, because a different species (possibly with a bridging carbyne-type ligand) can be formed[15] upon further acid attack on the precipitated carbyne complex.

Properties

The product is a green, crystalline, solid, unstable in air and in solution at ambient temperature under nitrogen (see below). Its infrared spectrum[2] (in Nujol mull) has bands at 3315 cm^{-1} (weak and broad), 2168 (strong), and 1524 cm^{-1} (strong), which are assigned to ν(NH), ν(C≡N) of the unreacted isocyanide, and ν(C═N) of the carbyne-type ligand, respectively. In the ^{13}C nmr spectrum in CD_2Cl_2, the carbyne-carbon, CNHMe, resonance appears at δ 242.4 ppm (weak and broad multiplet), downfield relative to $SiMe_4$. A singlet ^{31}P nmr resonance is observed[2] at 86.8 ppm, upfield from $P(OMe)_3$, in CD_2Cl_2, at $-42°$, whereas in the ^{1}H nmr spectrum[2] (in CD_2Cl_2, at $-40°$), the CN*H*Me resonance is a broad quartet at δ 3.50 ppm (relative to $SiMe_4$), and the CNH*Me* resonance is a doublet at 2.57 ppm [^2J(HNCH) = 4.7 Hz].

The following series of analogous carbyne complexes of tungsten and molybdenum may be synthesized similarly by reactions of *trans*-[$M(CNR)_2(dppe)_2$] (M = W; R = Me or Bu^t; M = Mo, R = Me) with various acids:[2] *trans*-[$M(CNHR)(CNR)(dppe)_2$]A [M = W; A = BF_4, HSO_4, SFO_3 or H_2PO_4 (for R = Me); A = BF_4, HSO_4 or SFO_3 (for R = Bu^t); M = Mo; A = BF_4, HSO_4 or FSO_3 (for R = Me)]. All of these products are green in color and exhibit bands assigned to ν(NH) (Nujol mull) in the 3318–3200 cm^{-1} region, to ν(C═N) of the carbyne-type ligand in the 1533–1515 cm^{-1} region (except for M = W, R = Bu^t, for which such a band is not observed), and to ν(C≡N) of the unreacted isocyanide in the 2180–2132 cm^{-1} region.[2] They always are contaminated with variable amounts of the corresponding hydride [$MH(CNR)_2(dppe)_2$]A complexes, which exhibit a strong, broad band, assigned to ν(C≡N), in the 2038–2000 cm^{-1} region.[3]

These monocarbyne-type complexes are deprotonated by base (carbonate), thus generating the parent diisocyanide complexes, and, in solution (e.g., in CH_2Cl_2), they undergo decomposition at ambient temperature with formation of the corresponding hydride complexes.[16] They are susceptible to further protonation, generating, in the solid state,[15] possible dimeric species with bridging carbyne-type ligands, and, in solution,[16] the dicarbyne-type *trans*-[$M(CNHR)_2(dppe)_2$]A_2 complexes (M = Mo or W, R = Me, see below).

C. *TRANS*-BIS[1,2-ETHANEDIYLBIS(DIPHENYLPHOSPHINE)]-BIS-[(METHYLAMINO)METHYLIDYNE]TUNGSTEN(IV) BIS(TETRAFLUOROBORATE)*

$$trans\text{-}[W(CNMe)_2(dppe)_2] + 2\ [Et_2OH][BF_4] \longrightarrow trans\text{-}[W(CNHMe)_2(dppe)_2][BF_4]_2 + 2\ Et_2O$$

*This complex may be considered a dicarbyne species if the carbyne-type ligands are regarded as having a formal (1-) charge, *i.e.*, MeHNC ⇉ W ⇇ $CNHMe^{2+}$. Alternatively, it may be considered as a dicarbene complex.

Procedure

A solution of *trans*-[W(CNMe)$_2$(dppe)$_2$] (1.55 g, 1.47 mmoles) in CH_2Cl_2 (270 mL) is placed, under nitrogen, into a 500-mL single-necked Schlenk flask. The solution is filtered under nitrogen into a similar Schlenk flask, and tetrafluoroboric acid (3.20 mmoles, 0.443 mL of commercial [Et_2OH][BF_4]) is rapidly (~15–30 sec) added dropwise, under nitrogen, to the vigorously stirred solution.

After ~0.5 hour, the solution is concentrated under vacuum to approximately one-third volume and diethyl ether is then added slowly and dropwise, under nitrogen, to the stirred solution to give a brownish suspension,* which is filtered, under nitrogen, into another Schlenk flask. Diethyl ether is again added slowly to the filtered and stirred solution, under nitrogen, until the dicarbyne product appears as a pink crystalline solid.† (Before this happens it may be necessary to extract additional portions of the brownish suspension, by the process outlined above, until precipitation is complete.) Cooling (e.g., to −20°) of the ethereal solution may also result in precipitation of the crystalline dicarbyne product. This complex is then filtered off, under nitrogen, washed with a (3:1) mixture of diethyl ether/dichloromethane, and dried under vacuum. The yield is 1.3 g (~1.0 mmole, ~68%). *Anal.* Calcd. for [W(CNHMe)$_2$(dppe)$_2$][BF_4]$_2$·$\frac{1}{2}$$Et_2O$ (red isomer): C, 54.6; H, 4.8; N, 2.2. Found: C, 54.9; H, 4.8; N, 2.5. Calcd. for [W(CNHMe)$_2$(dppe)$_2$[BF_4]$_2$·Et_2O (yellow isomer): C, 54.9; H, 4.9; N, 2.1. Found: C, 55.3; H, 5.2; N, 2.4.

A slow addition of acid should be avoided in this preparation, since a different complex (a hydrido-carbon species)[3] may then be the main product. Too great an excess of acid should also be avoided, due to the possible formation of a triprotonated product.[16] The method is general for the preparation of other analogous dicarbyne-type complexes of molybdenum and tungsten, using a variety of acids (see below).

Properties

The product is fairly air-stable in the solid state (for at least two weeks), but should be stored under nitrogen, especially if it is in solution. It is isolated as a mixture of two isomers, one red and the other yellow,‡ both with a strong band at 1630 cm^{-1} in their infrared spectra[2] (Nujol mull), which is assigned to ν(CN) of the carbyne-type ligands; sharp bands are also observed for ν(NH) at 3407 cm^{-1} or at 3422 cm^{-1} (yellow isomer). In solution, the yellow isomer appears to convert rapidly to the red one, which exhibits, in the ^{13}C nmr spectrum[2]

*This suspension appears to contain a heterogeneous mixture of other carbyne complexes.

†It is a mixture of two isomers (see below), one red and the other, yellow, which eventually may be separated by hand when distinct crystals are formed.

‡These isomers may correspond to a planar MeHNCWCNHMe system, with the NR and NH groups either *cis* or *trans* to each other.

(CD_2Cl_2 solution at −36°), a carbyne carbon (*C*NHMe) resonance (which is a broad multiplet) at δ 194.5 ppm downfield from $SiMe_4$. In the ^{31}P nmr spectrum,[2] two equal, intense resonances [at δ 103.1 and 104.3 ppm in CD_2Cl_2 at −55°, upfield from $P(OMe)_3$] are observed. In the ^{1}H nmr spectrum[2] (in CD_2Cl_2, at −35°), the CN*H*Me resonance is observed as a broad quartet at δ 4.25 ppm (relative to $SiMe_4$), whereas the CNH*Me* resonance is a doublet at δ 2.38 ppm [^{2}J(HNCH) = 4.2 Hz]. In the ^{19}F nmr spectrum[2] (in CD_2Cl_2 at 25°), only one resonance (at δ 151.4 ppm, upfield from $CFCl_3$) is observed in the −289 to +339 ppm region.

A series of analogous dicarbyne-type complexes of tungsten and molybdenum may be synthesized similarly,[2] from the reactions of the parent diisocyanide complexes with various acids: *trans*-$[M(CNHR)_2(dppe)_2]A_2$ [For M = W; R = Me; A = BF_4 (red and yellow isomers), HSO_4 (pink), SFO_3 (brownish), $SClO_3$ (brown). For M = W; R = C_6H_4Me-*4*; A = BF_4 (red). For M = Mo; R = Me; A = BF_4 (red and yellow isomers)]. They exhibit, in the infrared spectra[2] (in Nujol mull), bands assigned to ν(NH) in the 3422–3360 cm^{-1} region and to ν(CN) (of the carbyne-type ligands) in the 1645–1606 cm^{-1} region.

The parent diisocyanide complexes are regenerated upon deprotonation (by MeLi or NaOMe) of the dicarbyne-type complexes (R = Me). The ligating CNHMe group in the latter appears to be susceptible to electrophilic attack by acid (HBF_4) at the nitrogen, and to nucleophilic attack by hydride at the binding carbon, *trans*-$[W(CNH_2Me)(CNHMe)(dppe)_2][BF_4]_3$ and *trans*-$[Mo(CHNHMe)(CNHMe)(dppe)_2][BF_4]$ appearing to be formed, respectively.[16]

References

1. J. Chatt, C. M. Elson, A. J. L. Pombeiro, R. L. Richards, and G. H. D. Royston, *J. Chem. Soc. (Dalton)*, 165 (1978).
2. J. Chatt, A. J. L. Pombeiro, and R. L. Richards, *J. Chem. Soc. (Dalton)*, 492 (1980).
3. J. Chatt, A. J. L. Pombeiro, and R. L. Richards, *J. Chem Soc. (Dalton)*, 1585 (1979).
4. J. Chatt, A. J. L. Pombeiro, and R. L. Richards, *J. Organomet. Chem.*, **184,** 357 (1980).
5. A. J. L. Pombeiro, in *New Trends in the Chemistry of Nitrogen Fixation,* J. Chatt, L. M. C. Pina, and R. L. Richards (eds.), Academic Press, 1980, Ch. 10.
6. R. L. Richards, *Inorg. Syn.*, **19,** 174 (1979).
7. F. Bonati and G. Minghetti, *Inorg. Chim. Acta,* **9,** 95 (1974).
8. P. M. Treichel, *Adv. Organomet. Chem.*, **11,** 21 (1973).
9. I. Ugi, U. Fetzer, U. Eholzer, H. Knupfer, and K. Offermann, *Angew. Chem. Int. Ed.*, **4,** 472 (1965).
10. R. E. Schuster, J. E. Scott, and J. Casanova, *Org. Syn.*, **46,** 75 (1966).
11. J. Casanova, N. D. Werner, and A. E. Schuster, *J. Org. Chem.*, **31,** 3473 (1966).
12. J. R. Dilworth and R. L. Richards, *Inorg. Syn.*, **20,** 119 (1980).
13. A. J. L. Pombeiro and R. L. Richards, *J. Organomet. Chem.*, **179,** 459 (1979).
14. J. Chatt, K. W. Muir, A. J. L. Pombeiro, R. L. Richards, G. H. D. Royston, and R. Walker, *J. Chem. Soc. Chem. Commun.*, 708 (1975).
15. A. J. L. Pombeiro and R. L. Richards, *Rev. Port. Quim.*, **21,** 132 (1979).
16. A. J. L. Pombeiro and R. L. Richards, *Transition Met. Chem.*, **5,** 55 (1980).

4. (η^5-PENTAMETHYLCYCLOPENTADIENYL)COBALT COMPLEXES

Submitted by STEVEN A. FRITH and JOHN L. SPENCER*
Checked by W. E. GEIGER, JR. and J. EDWIN†

Pentamethylcyclopentadienyl complexes exhibit a chemistry complementary to that of the unsubstituted cyclopentadienyl analogs, with differences in behavior that may be ascribed to the greater bulk or the greater electron-releasing ability of the pentamethylcyclopentadienyl ligand.[1] The following syntheses lead to a group of complexes which have given access to rich areas of organo-cobalt chemistry. The syntheses have been designed such that later members of the group may be prepared without lengthy purification of the intermediates.

A. DICARBONYL-(η^5-PENTAMETHYLCYCLOPENTADIENYL)COBALT(I)

$$Co_2(CO)_8 + 2C_5Me_5H + C_6H_8 \longrightarrow 2[Co(\eta^5\text{-}C_5Me_5)(CO)_2] + C_6H_{10} + 4CO$$

Originally $[Co(C_5Me_5)(CO)_2]$ was prepared[2] by the photolysis of $Co_2(CO)_8$ and C_5Me_5H and subsequently[3] by the reaction of $Co_2(CO)_8$ and $C_5Me_5C(O)Me$ in refluxing cyclohexane. The method described here is a substantial modification of that used by Byers and Dahl[4] and gives excellent yields from both precursors.

Procedure

■ **Caution.** *Both dicobalt octacarbonyl and carbon monoxide are poisonous and therefore the reaction should be carried out in a well-ventilated hood.*

A 100-mL, two-necked, round-bottomed flask is fitted with a reflux condenser and a magnetic stirring bar. A T-junction, placed on top of the condenser, is connected to a supply of dry nitrogen and a mineral oil bubbler, and the flask is flushed for several minutes. Dry, deoxygenated dichloromethane (50 mL), octacarbonyldicobalt(0) (6.0 g, 17.5 mmoles), pentamethylcyclopentadiene‡ 3 g, 22.1 mmoles) and 1,3-cyclohexadiene (2.5 mL) are added through the second neck, which is then closed with a stopper. The mixture is stirred and heated to

*Department of Inorganic Chemistry, University of Bristol, Bristol BS8 1TS, United Kingdom.
†Department of Chemistry, University of Vermont, Burlington, VT 05405.
‡Available from Strem. Chem., P.O. Box 212, Danvers, MA 01923 or Aldrich Chemical Co., 940 W. St. Paul Ave., Milwaukee, WI 53233.

maintain a very gentle reflux. After 45 minutes, a further 2.3 g (16.9 mmoles) of pentamethylcyclopentadiene is added and heating is continued for another 1.75 hours. The reaction may be conveniently monitored, using IR spectroscopy, by the appearance of ν_{CO} bands at 1999 and 1935 cm^{-1} (CH_2Cl_2). When reaction is complete, there may also be weak absorptions due to traces of $Co_4(CO)_{12}$ (2065, 2055, and 1858 cm^{-1}) or $[Co(\eta^3\text{-}C_6H_9)(CO)_3]$ (2047 and 1983 (sh) cm^{-1}).

The condenser is replaced with a stopcock adaptor and the volatile components are evaporated at reduced pressure and ambient temperature. If the temperature is kept at or below 20°, the oily residue will crystallize spontaneously. This crude material may be used in the synthesis of $[CoI_2(\eta^5\text{-}C_5Me_5)(CO)]$, or it may be purified as follows. Nitrogen is readmitted to the flask, the crude crystalline mass is dissolved in dry deoxygenated hexane (30 mL), and the solution is applied to a column of alumina (Brockman, activity II) (20 cm × 2.5 cm), previously prepared under nitrogen and washed with deoxygenated hexane (200 mL). The product is eluted as an orange-brown band with hexane and collected under nitrogen. Evaporation of the solvent under reduced pressure yields deep-red crystals of $[Co(\eta^5\text{-}C_5Me_5)(CO)_2]$, 7.9–8.3 g (90–95% based on $Co_2(CO)_8$). Further purification may be effected by crystallization from hexane solution at low temperatures or by sublimation (40°, 0.01 torr). *Anal.* Calcd. for $C_{12}H_{15}O_2Co$: C, 57.61; H, 6.04. Found: C, 57.51; H, 6.26 (mp 58°, sealed tube).

Properties

Dicarbonyl(η^5-pentamethylcyclopentadienyl)cobalt(I) is a deep red-brown crystalline solid that may be handled briefly in air but should be stored in an inert atmosphere. It dissolves readily in dry, deoxygenated organic solvents to give stable solutions but, in the presence of air or water, it slowly decomposes. The ^1H nmr spectrum ($CDCl_3$) shows a sharp singlet at δ 1.98 ppm.

B. CARBONYLDIIODO-(η^5-PENTAMETHYLCYCLOPENTADIENYL)COBALT(III)

$$[Co(\eta\text{-}C_5Me_5)(CO)_2] + I_2 \longrightarrow [CoI_2(\eta\text{-}C_5Me_5)(CO)] + CO$$

The method described here is based on that of King and co-workers.[5] Roe and Maitlis[6] followed a similar procedure but used an alkane solvent.

Procedure

■ **Caution.** *A large volume of carbon monoxide is released in a short time. The reaction should be carried out in a well-vented hood.*

A 250-mL two-necked flask is fitted with a magnetic stirrer bar and a T-piece connected to a nitrogen supply and a mineral oil bubbler. The flask is purged with nitrogen, and sodium-dry diethyl ether (150 mL) is added, followed by 8.9 g of iodine (35.0 mmoles). The second neck is then closed with a rubber septum and a solution of $[Co(\eta^5\text{-}C_5Me_5)(CO)_2]$ (8.75 g, 35.0 mmoles, or all of the crude material from the previous synthesis) in deoxygenated diethyl ether (20 mL) is added dropwise from a syringe to the rapidly stirred reaction mixture. Addition may be completed in 5–10 minutes. A black, crystalline precipitate forms rapidly with the evolution of carbon monoxide gas. Stirring is continued for another hour. The mixture is filtered in the air using a sintered-glass funnel and the crude product is washed with diethyl ether (three × 20 mL portions) and dried in an oven at 80°. The material so-formed is sufficiently pure for the preparation of $[\{CoI_2(\eta^5\text{-}C_5Me_5)\}_2]$. However, it may be purified by dissolution in a large volume of dichloromethane to give a deep purple solution, which is filtered through a glass sinter (porosity 3) and evaporated to small volume using a rotary evaporator. Lustrous black crystals of $[CoI_2(\eta^5\text{-}C_5Me_5)(CO)]$ are filtered off and dried under vacuum. Yield: 15.0–15.8 g (90–95%). *Anal.* Calcd. for $C_{11}H_{15}OI_2Co$: C, 27.76; H, 3.18. Found: C, 27.70; H, 3.45.

Properties

Carbonyldiiodo(η^5-pentamethylcyclopentadienyl)cobalt(III) is a black, crystalline, air-stable solid. It dissolves readily in polar solvents forming deep purple solutions but is virtually insoluble in diethyl ether, alkanes, and aromatic solvents. The ^1H spectrum ($CDCl_3$) shows a singlet at δ 2.20 ppm. The IR spectrum exhibits a single metal-carbonyl stretching band at 2053 cm^{-1} (CH_2Cl_2). Removal of the iodide ligands by Ag^+ provides a facile route to cationic organometallic derivatives of Co(III).[7]

C. DI-μ-IODO-BIS[IODO(η^5-PENTAMETHYLCYCLOPENTADIENYL)COBALT(III)]

$$2[CoI_2(\eta^5\text{-}C_5Me_5)(CO)] \xrightarrow{\Delta} [\{CoI_2(\eta^5\text{-}C_5Me_5)\}_2] + 2\,CO + CoI_2$$

The method described here is based on that of Roe and Maitlis.[6] An analogous dichloro(η-ethyltetramethylcyclopentadienyl) derivative, $[\{CoCl_2(\eta^5\text{-}C_5Me_4Et)\}_2]$ has been prepared[8] by the extraction of $CoCl_2$ from $[Co_3Cl_6(\eta^5\text{-}C_5Me_4Et)_2]$ in turn made from $[Sn(n\text{-}Bu)_3(\sigma\text{-}C_5Me_4Et)]$ and $CoCl_2$.

Procedure

■ **Caution.** *The reaction releases toxic carbon monoxide and should be carried out in a well-ventilated hood.*

A 250-mL, two-necked, round-bottomed flask is charged with 150 mL of dry *n*-octane and 15.5 g of powdered [$CoI_2(\eta^5$-$C_5Me_5)(CO)$] and equipped with a magnetic stirring bar, a nitrogen inlet, and a reflux condenser. The top of the condenser is connected to a mineral oil bubbler venting into the hood. The rapidly stirred mixture is heated at reflux for 5 hours under a slow N_2 purge. During the reaction, the microcrystalline suspension is converted to a suspension of lustrous, black crystals. After cooling to room temperature, the crude [$\{CoI_2(\eta^5$-$C_5Me_5)\}_2$] is collected in a sintered-glass filtration funnel, washed with pentane, and dried in an oven at 80°. The product may be purified by dissolving it in a large volume of dichloromethane, filtering, and removing most of the solvent at reduced pressure on a rotary evaporator, or by using the continuous extraction apparatus shown in the figure. The crude material is transferred to the paper

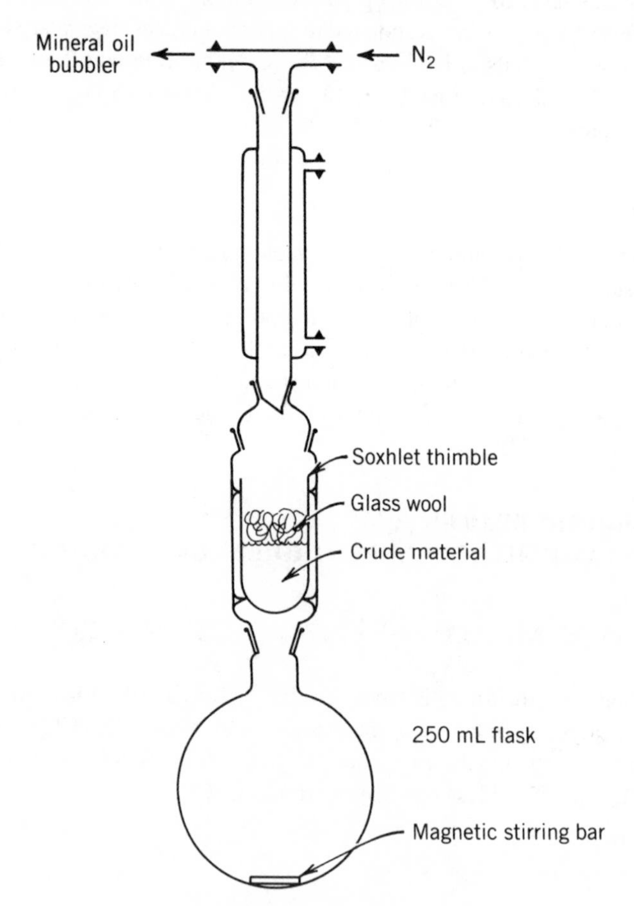

Continuous Extraction Apparatus

thimble, a loose plug of glass wool is packed above it, and dry dichloromethane (120 mL), contained in the flask, is boiled at a sufficient rate that condensate drips into the thimble. This process may be run conveniently overnight. The flask is then detached and cooled to 0° for three hours. The product, as lustrous, black crystals, is recovered by filtration and dried under vacuum (0.05 torr, 1 hr). Yield: 11.5 g. A further crop (~1.3 g) may be obtained by reducing the volume of the filtrate to 30 mL, but this material should be checked by IR to ensure that it contains no $[CoI_2(\eta^5\text{-}C_5Me_5)(CO)]$. Total yield of $[\{CoI_2(\eta^5\text{-}C_5Me_5)\}_2]$ is 12.8 g (88%). *Anal.* Calcd. for $C_{20}H_{30}I_4Co_2$: C, 26.81; H, 3.38. Found: C, 26.71; H, 3.50.

Properties

Di-μ-iodo-bis[iodo(η^5-pentamethylcyclopentadienyl)cobalt(III)] is a black, crystalline, air-stable solid. It dissolves in polar solvents, forming dark green solutions, but is virtually insoluble in diethyl ether, alkanes, and aromatic solvents. The ^1H nmr spectrum ($CDCl_3$) exhibits a singlet at δ 1.80 ppm.

D. BIS(η^2-ETHENE)-(η^5-PENTAMETHYLCYCLOPENTADIENYL)COBALT(I)

$$[\{CoI_2(\eta^5\text{-}C_5Me_5)\}_2] + 4Na + 4C_2H_4 \longrightarrow 2[Co(\eta^5\text{-}C_5Me_5)(\eta^2\text{-}C_2H_4)_2] + 4NaI$$

Green and Pardy[7] prepared $[Co(\eta^5\text{-}C_5Me_4Et)(\eta^2\text{-}C_2H_4)_2]$ by the sodium-amalgam reduction of $[\{CoCl_2(\eta^5\text{-}C_5Me_4Et)\}]$ under ethylene (toluene, 110°). The pentamethylcyclopentadienyl complex described here offers several advantages: it is a crystalline solid rather than an oil; compounds derived from $[Co(\eta^5\text{-}C_5Me_5)(\eta^2\text{-}C_2H_4)_2]$ by displacement of ethylene are frequently solids also; and the ^1H nmr is simpler.

Procedure

■ **Caution.** *Mercury vapor is a cumulative poison and the large volumes of ethylene used in this synthesis are a fire hazard. The preparation should be carried out in a well-ventilated hood in the absence of naked flames. Drying tetrahydrofuran is hazardous and should be carried out with due caution.*[9]

A 250-mL amalgam reaction vessel,[10] fitted with a powerful electrically driven paddle stirrer, is flushed with nitrogen and charged with 20 mL of clean mercury. Sodium (2.8 g) is cleaned and finely cut under hexane, and is added piece by piece to the reaction flask. The nitrogen source is replaced by a source of ethylene, and 150 mL of dry, deoxygenated tetrahydrofuran is added through the third neck. Stirring is continued as the flask is flushed with ethylene. Pure* solid $[\{CoI_2(\eta^5\text{-}C_5Me_5)\}_2]$ (10.0 g, 11.15 mmoles) is added through the third neck, which is then closed, and the mixture is beaten very rapidly. As soon as the dark green color changes to mid-brown (~3–5 min), stirring is halted and the amalgam is drained from the reactor. A syringe is used to transfer the solution to a N_2-filled 250-mL, three-necked flask connected by stopcock adaptors to nitrogen and vacuum. Solvent is evaporated at reduced pressure and nitrogen is admitted to the flask. The residue is dissolved in 40 mL of dry, deoxygenated hexane and the solution is filtered under nitrogen through a 1-cm pad of Kieselguhr. The red-brown filtrate is reduced in volume to 20 mL by evaporation at reduced pressure, and the product is crystallized at −78° under nitrogen. The supernatant liquid is decanted with a syringe and the orange-brown crystals are dried at 20° and 0.05 torr for 1 hour. *Anal.* Calcd. for $C_{14}H_{23}Co$: C, 67.20; H, 9.80. Found: C, 67.70; H, 10.04 (mp 77–78°; sealed tube).

A second crop of crystals may be obtained from the mother liquor by filtering, evaporating, and cooling as before. Combined yield: 3.9–4.5 g (70–85%).

Properties

Bis(η^2-ethylene)(η^5-pentamethylcyclopentadienyl)cobalt(I) is an orange-brown, volatile, crystalline solid. It may be handled briefly in air but if it is to be kept for long periods, it should be under an inert atmosphere at −20°. It is soluble in organic solvents but decomposes in the presence of air and water. In coordinating solvents, decomposition occurs slowly; therefore, it is best to allow it to react in alkane, diethyl ether, or aromatic solvents. The ^1H nmr spectrum (C_6D_6) consists of a singlet at δ 1.43 [15 H, C_5Me_5] and an $|AB|_2$ pattern for the C_2H_4 protons, δ_A 1.60, δ_B 0.86 ppm ($|J(AB) + J(AB')|$ 13 Hz). The IR spectrum (Nujol) has strong bands at 1197 and 1180 cm^{-1}.

References

1. (a) R. B. King, *Coord. Chem. Rev.*, **20,** 155 (1976); (b) P. M. Maitlis, *Chem. Soc. Rev.*, **10,** 1 (1981); (c) T. P. Wolczanski and J. E. Bercaw, *Accts. Chem. Res.*, **13,** 121 (1980); (d) J. M. Manriquez *et al.*, *Inorg. Syn.*, **21,** 181 (1982).
2. R. B. King and M. B. Bisnette, *J. Organomet. Chem.*, **8,** 287 (1967).
3. R. B. King and A. Efraty, *J. Am. Chem. Soc.*, **94,** 3773 (1972).
4. L. R. Byers and L. F. Dahl, *Inorg. Chem.*, **19,** 277 (1980).
5. R. B. King, A. Efraty, and W. M. Douglas, *J. Organomet. Chem.*, **56,** 345 (1973).

*If $[CoI_2(\eta^5\text{-}C_5Me_5)(CO)]$ is present, reduction will give the intense green compound $[\{Co(\eta\text{-}C_5Me_5)(\mu\text{-}CO)\}_2]$, which will mask the progress of the reaction and be difficult to separate later.

6. D. M. Roe and P. M. Maitlis, *J. Chem. Soc. (A)*, 3173 (1971).
7. G. Fairhurst and C. White, *J. Chem. Soc., Dalton Trans.*, 1524, 1531 (1979).
8. M. L. H. Green and R. B. A. Pardy, *J. Chem. Soc., Dalton Trans.*, 355 (1979).
9. R. W. Parry, *Inorg. Syn.*, **12,** 317 (1970).
10. B. D. Dombek and R. J. Angelici, *Inorg. Syn.*, **17,** 100 (1977).

5. DIBROMODIPHENYLTIN(IV)

Submitted by R. K. CHANDRASEKARAN and JOHNSON D. KOOLA*
Checked by D. W. GOEBEL, JR. and J. P. OLIVER†

$$(acac)_2SnX_2 + 2\ RMgX \longrightarrow R_2SnX_2 + 2\ acacMgX$$

The laboratory procedure for the synthesis of dibromodiphenyltin and other dihalodiorganotin compounds involves first the preparation of tetraorganotin compounds, R_4Sn, and then disproportionation with tin(IV) halides, SnX_4, at higher temperatures.[1] Although well established, this method is very time-consuming. The dihalodiorganotin compounds can be obtained by a simple and convenient one-step process[2] involving the reaction of a Grignard reagent with $(acac)_2SnX_2$ (acacH = 2,4-pentanedione; X = Cl, Br).

The reaction is mildly exothermic and is about 75% complete at room temperature. However, refluxing is carried out to ensure complete reaction.

Experiments designed to prepare 8–10 g of the end product can be finished in 8–9 hours. To prepare 40–50 g of the end product requires two days, since the removal of precipitated acacMgX by filtration is slow. Employing $(acac)_2SnX_2$ as the starting material, any R_2SnX_2 compound can be synthesized by treatment with the appropriate Grignard reagent.

Dihalodiorganotin compounds are useful as polyvinyl chloride stabilizers,[3] fungicides, insecticides,[4] and key intermediates for the syntheses of other organotin compounds.

Procedure

■ **Caution.** *Organotin compounds are very toxic and should be handled accordingly.*

All operations are performed in a nitrogen atmosphere, and syringes are used for the transfer of solvents and solutions. Bromophenylmagnesium is prepared from 6 g (0.25 mole) of magnesium and 35 g (0.223 mole) of bromobenzene

*Department of Inorganic Chemistry, University of Madras, Guindy, Madras-600 025, India.
†Department of Chemistry, Wayne State University, Detroit, MI 48202.

in 200 mL of dry diethyl ether. To the Grignard solution, dry benzene (75 mL) is added to slow the easy evaporation of diethyl ether. The Grignard content per mL is determined by titration. The $(acac)_2SnBr_2$ is prepared by a reported method.[5]

The $(acac)_2SnBr_2$ (47.6 g, 0.1 mole) is placed in a 1-L round-bottomed flask fitted with a dropping funnel and a reflux condenser, and is dissolved in 400 mL of dry benzene. Bromophenylmagnesium (37.5 g, 0.207 mole) is added dropwise from the dropping funnel. After the addition is complete, the reaction contents are refluxed in an oil bath for three hours. The precipitated acacMgBr is filtered off using sintered-glass filter, in a nitrogen atmosphere, and the solvents are removed from the filtrate by distillation. If a solid residue is observed in the distillation flask, the product is filtered again and washed with the minimum quantity of dry benzene. The filtrate is again distilled to remove benzene. The product is finally obtained by vacuum distillation (bp 160°–165° at 5×10^{-1} torr). Yield: 26.5 g; 61%. Very pure dibromodiphenyltin is obtained by redistillation. The product is characterized by nmr and mass spectrometry and by the formation of the bipyridyl adduct.[6]

Properties

The colorless solid is reported by the checkers to melt at 41–42° (Lit. 42°).

References

1. K. A. Kozeschkow, *Ber.*, **62,** 996 (1929); **66,** 1661 (1933).
2. R. K. Chandrasekaran, S. Venkataraman, and Johnson D. Koola, *J. Organomet. Chem.*, **215,** C_{43}–C_{44} (1981).
3. A. M. Frye, R. W. Horst, and M. A. Paliobagis, *J. Polymer Sci. A*, **2,** 1765 (1964).
4. R. C. Poller, *The Chemistry of Organotin Compounds,* Logos Press, London, 1970, p. 274.
5. G. T. Morgan and H. D. K. Drew, *J. Chem. Soc.*, 372 (1924).
6. W. R. McWhinnie, R. C. Poller, J. N. R. Ruddick, and M. Thevarasa, *J. Chem. Soc.* (*A*), 2327 (1969).

6. CARBONYLPENTAFLUOROPHENYL COBALT(I) AND (II) COMPLEXES

Submitted by PASCUAL ROYO,* NATIVIDAD ESPAÑA,* and AMELIO VÁZQUEZ DE MIGUEL*

Checked by T. J. GROSHENS and K. J. KLABUNDE†

Both organocobalt(I) and (II) carbonyl complexes can be synthesized from tetrahydrofuran (THF) solutions of bis(pentafluorophenyl)cobalt(II), and this is an

*Universidad de Alcalá de Henares, Spain.

†Department of Chemistry, Kansas State University, Manhattan, KS 66506.

excellent way to isolate different types of phosphine carbonyl cobalt complexes.

Tetrahydrofuran solutions of bis(pentafluorophenyl)cobalt(II) have been prepared[1] from anhydrous cobalt halides and the Grignard reagent. These solutions react with different phosphines[1,2] to give bis(pentafluorophenyl)bis-(phosphine)cobalt(II) derivatives. The reaction of these complexes with CO at atmospheric pressure gives pentacoordinate carbonyl cobalt(II) compounds.[2] Reaction of the organocobalt(II) THF solution with CO can also cause reduction of tetracarbonyl-pentafluorophenylcobalt(I),[3] which is obtained in a higher yield than by previous methods.[4] This compound reacts with different amounts of various phosphines (L) to give $Co(C_6F_5)(CO)_3L$[3] and $Co(C_6F_5)(CO)_2L_2$.[5]

A. CARBONYL COBALT(I) COMPLEXES

1. Tetracarbonylpentafluorophenylcobalt(I)

$$CoBr_2 + 2\ Mg(C_6F_5)Br \longrightarrow Co(C_6F_5)_2 + 2\ MgBr_2$$

$$Co(C_6F_5)_2 + 4CO \longrightarrow Co(C_6F_5)(CO)_4 + \text{redox products}$$

Procedure

■ **Caution.** *This synthesis should be performed in a very efficient hood and the pressure bottle should be adequately shielded. Carbon monoxide is a highly toxic, colorless, odorless gas, and precautions must be taken to avoid exposure to this gas.*

Anhydrous cobalt(II) bromide (21.9 g, 100.0 mmoles), freshly distilled THF (150 mL), and a magnetic, Teflon-coated stirring bar are placed in a 1-L round-bottomed, two-necked flask, fitted with a rubber stopper and a condenser topped with a stopcock that is connected to nitrogen and to a vacuum line. The flask is alternatively evacuated and filled with dry nitrogen three times.

A THF solution (250 mL) of bromo-pentafluorophenylmagnesium, obtained[6] from magnesium (5.50 g, 230.0 mmoles) and bromopentafluorobenzene (25 mL, 200.0 mmoles) is transferred to the flask with stirring under nitrogen pressure. After addition, the dark blue solution is refluxed, with stirring, for two hours, and is then cooled to room temperature. Dioxane (60 mL) is slowly added by means of a syringe (with stirring). The flask is then closed with rubber stoppers, and is allowed to stand overnight at $-10°$. The decanted solution is filtered under nitrogen through a Celite filter-aid to a 2-L two-necked, round-bottomed flask equipped with a stopcock connected to the nitrogen and vacuum lines. After replacing the filter by a rubber stopper, the solution is evaporated to dryness under reduced pressure. The dark blue residue is washed four times with 50 mL of hexane, added with a syringe.

One liter of distilled dry hexane is added (with stirring) to obtain a blue suspension, and the rubber stopper is replaced by a gas inlet tube connected to the CO pressure bottle. Carbon monoxide is passed slowly through the solution by way of the gas inlet tube (with stirring) for six hours, and stirring is continued for an additional two hours, to give a green-yellow solution and a small amount of a blue solid. The solution is filtered under nitrogen pressure into a two-necked flask equipped with a rubber stopper and connected to nitrogen and vacuum lines through a stopcock.

The yellow solution (cooled to $-10°$) is evaporated to dryness to give crude $Co(C_6F_5)(CO)_4$ (18.2 g, 53.8 mmoles. Yield: 82%.* This compound can be purified by sublimation at 20–25° under reduced pressure. *Anal.* Calcd. for $Co(C_6F_5)(CO)_4$: C, 35.53; Co, 17.43. Found: C, 35.89; Co, 17.06.

Properties

The complex is a yellow, crystalline solid soluble in organic solvents, including hexane. The solid is unstable and decomposes slowly at room temperature and rapidly in solution. It can be stored indefinitely at $-20°$ under nitrogen. Melting point: 36°. The infrared spectrum shows ν(CO) at 2125(s), 2060(vs), and 2040(vs) (hexane). The complex serves as a useful reagent for the preparation of carbonyl phosphine cobalt(I) complexes, as one or two CO groups can be readily displaced at room temperature by different ligands.

2. Tricarbonylpentafluorophenyltriphenylphosphinecobalt(I)

A two-necked 250-mL round-bottomed flask equipped with a magnetic stirring bar, a stopcock connected to nitrogen and vacuum lines, and a rubber stopper is charged with solid $Co(C_6F_5)(CO)_4$ (1.50 g, 4.4 mmoles) while flushing with nitrogen. The crude compound can be used in this reaction, although is better to use the complex previously purified by sublimation. This is especially true when reactions with other phosphines are carried out, since the higher solubility of complexes makes their final purification difficult if an excess of phosphine is present. After the flask is evacuated and filled with nitrogen, freshly distilled, dry hexane (50 mL) is transferred into it under nitrogen. The rubber stopper is replaced by a pressure-equalizing dropping funnel topped with a stopcock that is connected to the nitrogen-vacuum lines. The funnel is charged with a solution of PPh_3 (1.164 g, 4.4 mmoles) in hexane (20 mL).

The addition of this solution immediately produces a precipitate of a yellow solid. The progress of the reaction may be monitored by following the changes in the ν(CO) region of the infrared spectrum of the solution. When the reaction is complete, the absorption bands at 2125 and 2040 cm^{-1} disappear completely.

*The checkers report 21%.

The reaction is complete after 40 minutes of stirring at room temperature. The solvent is eliminated by filtration through a filter under nitrogen. The solid is dried under vacuum. The compounds can be recrystallized by evaporation of a benzene-hexane solution to give yellow-orange crystals, which are filtered and dried under vacuum. (1.55 g, 2.7 mmoles. Yield: 62%). *Anal.* Calcd. for $Co(C_6F_5)(CO)_3(PPh_3)$: C, 56.66; H, 2.64; Co, 10.29. Found: C, 56.43; H, 2.56; Co, 10.00.

Properties

The complex is a yellow, crystalline solid soluble in most organic solvents, such as benzene and chloroform, but insoluble in hexane. It is stable indefinitely in the solid state under nitrogen at room temperature. The IR spectrum shows $\nu(CO)$ absorptions at 1980(vs) and 2060(m) cm^{-1} in Nujol, as well as characteristic C_6F_5 absorptions at 1490–1500(s), 1050–1035(s) and 955(vs).

Similar reactions can also be carried out with triethylphosphine and tributylphosphine, and their complexes are much more soluble in hexane. It is important to use pure $Co(C_6F_5)(CO)_4$ and to monitor the progress of the reaction by the infrared spectrum of samples of the resulting solution (taken by syringe). The hexane solutions obtained after filtration are crystallized by evaporating and cooling to $-78°$.

3. Dicarbonylpentafluorophenylbis(triphenylphosphine)cobalt(I)

A two-necked, 250-mL flask equipped with a magnetic stirring bar, a stopcock connected to nitrogen and vacuum lines, and a rubber stopper is charged with $Co(C_6F_5)(CO)_4$ (1.0 g, 2.95 mmoles) under a continuous flow of nitrogen, and is then evacuated and filled with nitrogen several times. Freshly distilled dry benzene (50 mL) is transferred into the flask under nitrogen. The rubber stopper is replaced by a pressure-equalizing dropping funnel topped with a rubber stopper. The funnel is charged with a solution of PPh_3 (1.552 g, 5.9 mmoles) in benzene (20 mL). This solution is added slowly to the flask with continuous stirring, producing a reddish solution.

The progress of the reaction can be monitored by following changes in the $\nu(CO)$ region. The characteristic absorptions of the starting compound change to give those of the monophosphine complex described above, and finally a new absorption at 1900 cm^{-1} appears while the 2060 cm^{-1} band disappears at the same time. The reaction is complete after 3.5 hours of stirring at room temperature. This time can be reduced by mild warming in a water bath at 40°. The resulting red solution is evaporated to dryness under vacuum to give a pale orange solid. This solid can be recrystallized from benzene-hexane by evaporation of the solvent and cooling the solution to give 1.5 g of product. Yield:

63%. *Anal.* Calcd. for $Co(C_6F_5)(CO)_2(PPh_3)_2$: C, 66.01; H, 3.90; Co, 7.31. Found: C, 65.52; H, 3.75; Co, 7.25.

Properties

The complex is an orange, crystalline solid soluble in most organic solvents, such as benzene and chloroform, but insoluble in hexane. It is stable at room temperature in the solid state under nitrogen. The IR spectrum shows $\nu(CO)$ absorptions at 1980 and 1900 cm^{-1} in Nujol (1960 and 1915 cm^{-1} in hexane solution), and characteristic C_6F_5 bands are present at 1490–1500(s), 1050–1035(s), and 955(vs).

Complexes with triethylphosphine and trin-butylphosphine can be isolated by following a similar procedure, using hexane as solvent, since these complexes are much more soluble.

B. CARBONYL COBALT(II) COMPLEXES

1. Carbonyl-bis(pentafluorophenyl)-bis(triethylphosphine)cobalt(II)

A two-necked, 250-mL flask equipped with a stopcock connected to nitrogen and vacuum lines and a rubber stopper is charged under flowing nitrogen with 9.7 g (15.4 mmoles) of the dark blue solid obtained in Section A-1, after it is washed with hexane and dried under vacuum. The flask is then evacuated and filled with dry nitrogen several times. Freshly distilled dry benzene (150 mL) is transferred into the flask under nitrogen. Triethylphosphine (4.5 mL, 30.5 mmoles) is added by syringe and the rubber stopper is replaced by a gas inlet tube connected to the CO pressure bottle. Carbon monoxide is bubbled slowly through the solution with stirring for 0.5 hour. The resulting dark green solution is evaporated under vacuum to give dark green crystals of the carbonyl complex. The solid is filtered, washed with hexane, and dried under reduced pressure. Yield: 6 g (60%). *Anal.* Calcd. for $Co(C_6F_5)_2(CO)(PEt_3)_2$: C, 45.68; H, 4.60; Co, 8.96. Found: C, 45.63; H, 4.47; Co, 8.73.

Similar reactions can be carried out with tributylphosphine and ethyldiphenylphosphine. The reaction with triphenylphosphine must be carried out with hexane instead of benzene. In this case, the unreacted solid is separated by filtration, and the solution is treated as described above.

Properties

These carbonyl complexes are stable indefinitely under nitrogen, but evolve CO in air at room temperature, this decomposition being complete for all complexes except $Co(C_6F_5)(CO)(PEt_3)_2$, which decomposes partially after several weeks. The same decomposition is more rapid with heating, and is complete at 110–120° for all complexes. All the complexes are paramagnetic with a μ(eff) of 2.2–2.4 BM at room temperature. The IR spectrum shows ν(CO) absorptions at 1990(vs) (PEt_3); 1995(vs) ($Pn\text{-}But_3$); 2020(vs) (PPh_3); 2019(vs) (dPPe) in benzene.

References

1. C. F. Smith and C. Tamborski, *J. Organomet. Chem.*, **32,** 257 (1971).
2. P. Royo and A. Vázquez, *J. Organomet. Chem.*, **204,** 243 (1981).
3. P. Royo and A. Vázquez, *J. Organomet. Chem.* **205,** 223 (1981).
4. J. B. Wilford, A. Forster, and F. G. A. Stone, *J. Chem. Soc.*, 6519 (1965).
5. N. España, P. Gomez, P. Royo, and A. Vázquez de Miguel, *J. Organomet. Chem.*, **256,** 141 (1983).
6. E. Nield, R. Stephens, and J. C. Tatlow, *J. Chem. Soc.*, 3227 (1962).

7. POTASSIUM μ-HYDRIDO-BIS[PENTACARBONYLCHROMATE(0)] AND POTASSIUM μ-HYDRIDO-BIS[PENTACARBONYLTUNGSTATE(0)]

$$2\,M(CO)_6 + 4OH^- \longrightarrow [M_2(CO)_{10}(\mu\text{-}H)]^- + CO_3^{2-} + H_2O + HCO_2^-$$

Submitted by M. D. GRILLONE*
Checked by C. OVALLES† and D. J. DARENSBOURG†

The $[M_2(CO)_{10}(\mu\text{-}H)]^-$ species (M = Cr, Mo, or W) were first isolated as tetramethylammonium salts by water treatment of the $[M_2(CO)_{10}]^{2-}$ anions, produced by the reaction of $M(CO)_6$ with $NaBH_4$ in liquid ammonia.[1] Later, they were obtained as tetraethylammonium salts by the reaction of $M(CO)_6$ with tetraethylammonium tetrahydroborate(1−) in diglyme, 1,1′-oxybis(2-methoxyethane),[2] and by the more convenient reaction of $M(CO)_6$ with $NaBH_4$ in tetrahydrofuran, followed by treatment with Et_4NBr.[3]

The title compound, $K[Cr_2H(CO)_{10}]$, has been prepared in about 80% yield by the reaction of $Cr(CO)_6$ with KOH in a methanol-tetrahydrofuran (THF) mixture containing some water.[4] From this compound, the tetrabutylammonium ($n\text{-}Bu_4N^+$) and μ-nitrido-bis(triphenylphosphine)(1+) $(Ph_3P)_2N^+$, (PPN^+) salts have been prepared by metathesis.[4,5]

*Institute of Inorganic Chemistry, University of Palermo, Palermo, 90128 Italy.
†Department of Chemistry, Texas A&M University, College Station, TX 77843.

The synthesis reported here is a modification of the original method. It utilizes different conditions ($[Cr(CO)_6]$:KOH molar ratios, solvent ratios (v/v), etc.), which make it possible to obtain the $[Cr_2(CO)_{10}(\mu\text{-}H)]^-$ species directly. The procedure requires 9–10 hours with 85–90% yields.

The compound $K[W_2H(CO)_{10}]$ can also be prepared by the reaction of $W(CO)_6$ with KOH, using a MeOH-THF mixture containing a minimum of water. Its synthesis is described here in detail. The procedure requires 25–27 hours with 45–55% yields.

The $[M_2(CO)_{10}(\mu\text{-}H)]^-$ derivatives have been the subject of several recent structural studies.[6] X-ray and infrared studies on its tetraethylammonium salt, $[Cr_2(CO)_{10}(\mu\text{-}H)]^-$, first indicated this to be the first example of a species containing a linear electron-deficient M-H-M system. This was later found to be nonlinear by neutron diffraction studies.[7] The chemical reactivity of these complex species is currently being studied, because they are of interest both for their applications in catalysis[8] and for the preparation of other metal carbonyls. The $[Cr_2(CO)_{10}(\mu\text{-}H)]^-$ complex, for example, is easily converted into $[Cr_2(CO)_{10}(\mu\text{-}X)]^-$ or into $[Cr(CO)_5X]^-$, thus providing a simple and rapid method for a high-yield synthesis of these species.[5]

A. POTASSIUM μ-HYDRIDO-BIS[PENTACARBONYLCHROMIUM(O)]

Procedure

All the operations must be carried out under a very pure nitrogen atmosphere ($O_2 \leq 2$ ppm V), using a high vacuum/nitrogen apparatus. Distilled water and solvents are carefully deaerated under vacuum and saturated with nitrogen before use. These latter and KOH may be laboratory grade. The $Cr(CO)_6$ is obtained commercially (Fluka or Merck), and is purified by sublimation in static vacuum at 60–70° before use.

■ **Caution.** *All metal carbonyls are very toxic compounds. They should be handled with extreme care in well-ventilated fume hoods.*

A solution of KOH (2.24 g, 40 mmoles) in water (1.8 mL) is prepared in a 250-mL, two-necked, round-bottomed Pyrex flask, equipped with a stopcock in the narrow neck (for nitrogen inlet and evacuation of the system), a magnetic stirring bar, and a rubber stopper. This solution is degassed with stirring under vacuum and saturated with nitrogen three times. Then, while a good nitrogen stream is passed through it, MeOH (18 mL) and then THF (36 mL), degassed and under nitrogen, are added. The solution is again degassed and saturated with nitrogen; then $Cr(CO)_6$ (2.2 g, 10 mmoles) is added to it. The resulting mixture is warmed under nitrogen with stirring in a water bath that is rapidly heated to

52–54°, and maintained at this temperature for three hours. During this time, a pale-yellow solid separates, and the solution becomes yellow-orange, and any $Cr(CO)_6$ and/or reaction intermediate is completely converted into the $[Cr_2(CO)_{10}(\mu\text{-}H)]^-$ species. The reaction mixture is left to stir under nitrogen for about four hours in the water bath, and is allowed to cool to room temperature. It is then stirred for about 30 minutes in a water–ice bath. Under a good nitrogen stream, water (50 mL), well-deaerated and under nitrogen, is gradually added by means of a syringe to this stirred mixture. The resulting cloudy solution is evaporated under vacuum, with stirring, in a water bath at 35–40° to remove the nonaqueous solvents (about 54 mL, collected in the usual trap in an acetone–dry-ice bath). The yellow, crystalline product separates almost quantitatively.

The resulting aqueous mixture is allowed to stir under nitrogen in a water–ice bath for about 30 minutes. Then the solid is collected by filtration into a cylindrical filter with a sintered-glass disk of G2 porosity and with a side-arm with stopcock (for nitrogen inlet and vacuum drying of the product), washed rapidly with water (three × 3 mL), precooled in a water–ice bath, and dried under high vacuum (10^{-3} torr) for several hours. Yield 1.92 g, 90.5%. The product is sufficiently pure for most purposes. The infrared ν_{CO} region of its Nujol mull confirms the absence of any $Cr(CO)_6$ resulting from decomposition (sharp band at about 1880 cm^{-1}).

It can be recrystallized from water. A sample (1 g) is dissolved in about 20 mL of water that is deaerated and under nitrogen, and is warmed (2–3 min.) with stirring in a water bath preheated to 54–55°. After rapid filtration under a slightly positive pressure of nitrogen, using a filter vessel with a sintered-glass disk of G4 porosity, the solution is allowed to sit for about five hours at room temperature, and then for about two hours in a water–ice bath. (Prolonged standing of this solution, even if under the usual slow nitrogen stream, may cause decomposition to some extent.) The crystals are collected in a cylindrical filter, like the one described before (G2 frit), washed rapidly with precooled water (two × 1 mL), and well-dried under high vacuum. Usual yield, 70–75%. The synthesis can be carried out on a larger scale (4–5 g of $Cr(CO)_6$) or on smaller scale (1–1.2 g of $Cr(CO)_6$) without appreciable differences.

B. POTASSIUM μ-HYDRIDO-BIS[PENTACARBONYLTUNGSTEN(O)]

Procedure

As in the previously-described procedure, all the operations must be carried out under nitrogen ($O_2 \leq 2$ ppm by volume), using a high-vacuum nitrogen apparatus. Distilled water and solvents are carefully deaerated under vacuum and saturated with nitrogen before use. These solvents and KOH are reagent grade. Commercial $W(CO)_6$ (Fluka or Merck) is purified by sublimation in static vacuum at about 70° before use.

A solution of KOH (3.94 g, 70 mmoles) in water (1.8 mL) is prepared in a two-necked, 250-mL round-bottomed Pyrex flask, equipped with a stopcock on the side narrow neck (inlet for nitrogen or vacuum), a magnetic stirring bar, and a rubber stopper. This solution is deaerated under vacuum with stirring, and is saturated with nitrogen three times. Then, while a good nitrogen stream is passed through it, MeOH (25 mL) and THF (50 mL), degassed and under nitrogen, are added. The solution is again degassed and saturated with nitrogen. Then $W(CO)_6$ (2.46 g, 7 mmole) is added to it. The resulting mixture is warmed under nitrogen, with stirring, in a water bath that is rapidly heated to 52–55°, and maintained at this temperature for 19 hours. During this time a yellow solid separates from the pale-yellow solution. The reaction mixture is left to stir under nitrogen for about five hours in the water bath, and is allowed to cool to room temperature. Then, it is allowed to cool for about 30 minutes in a water–ice bath. To this stirred, cool mixture, water (70 mL), degassed and under nitrogen, precooled in a water–ice bath, is gradually added by means of a syringe. (Rapid addition can induce gas evolution with concomitant $W(CO)_6$ formation, which leads to significant losses due to decomposition. Care should be taken to avoid warming.) The resulting cloudy solution is partly evaporated under vacuum, with stirring, in a water bath at 25–27°, to remove the nonaqueous solvents (about 70 mL in the trap consisting of dry ice and acetone), which permits an almost quantitative separation of the pale-yellow reaction product. (Some decomposition can occur during evaporation because of warming and/or prolonged reaction times. A water–bath temperature between 35–40° can be used only to remove the last traces of the solvents more rapidly.) The aqueous mixture is saturated with nitrogen, and allowed to stir under nitrogen for about 15 minutes in a water–ice bath. Then, the crystalline solid is collected by filtration into a cylindrical filter with a sintered-glass disk of G3 porosity and with a side stopcock for nitrogen inlet and vacuum drying of the product, rapidly washed with precooled water deaerated and under nitrogen (three × 1 mL), and dried under high vacuum (10^{-3} torr) at room temperature for several hours.* Yield: 1.16–1.3 g, 48–55%.

The product can be recrystallized from water. A sample (1 g) is dissolved under nitrogen in 18 mL of water with brief warming (3 min.) in a water bath preheated to 52–55°. After rapid filtration under a slight positive pressure of nitrogen, through a vessel with a sintered-glass disk of G4 porosity, the solution is allowed to sit under nitrogen for three hours at room temperature, and then

*The absence of any $W(CO)_6$, due to decomposition, should be checked by infrared control in the carbonyl region of Nujol mull (sharp band at about 1875 cm^{-1}). It can be removed by prolonged vacuum treatment.

for about two hours in a water–ice bath. The crystals are collected under nitrogen onto a G2 frit like that described before, washed with precooled water (two × 1 mL), and well-dried under high vacuum. Usual yield, 68–73%. *Anal.* Calcd. for $C_{10}HW_2KO_{10}$: C, 17.46; H, 0.15; W, 53.45; K, 5.68. Found: C, 17.53; H, 0.21; W, 53.5; K, 5.52.

The PPN* or Bu_4N* salts are prepared by allowing a sample of the potassium salt (1 mmole) and of PPNCl* (1.2 mmoles) or Bu_4NBr, to react under nitrogen in anhydrous acetone (~30 mL) at room temperature for about one hour. After filtration of KCl or KBr (G4 frit), the filtrate is evaporated to dryness under vacuum in a water bath at 35–37°. The resulting yellow, sticky residue is dissolved in 96% ethanol (about 50 mL or 15 mL [for KBr]) with brief warming (three min.) in a pre-heated water bath at 52–54°. After filtration through a G4 frit, water, deaerated and under nitrogen, is added almost dropwise to the filtrate, and the mixture is manually stirred, until a persistent turbidity appears, which is removed after further very short warming in a pre-heated water bath at 55–57°. The solution is left to sit under nitrogen at room temperature for about three hours, and then in a water–ice bath for about two hours. The crystals are collected under nitrogen onto a G2 frit, washed with an ethanol–water solution (1:3), and then well-dried under high vacuum. By concentration under vacuum of the filtrate and washing, additional crops of crystals are obtained. Usual yield, 75–85%.

Properties

The compounds $K[Cr_2H(CO)_{10}]$ and $K[W_2H(CO)_{10}]$ are yellow-orange and pale-yellow crystalline solids, respectively, and are moderately air-stable but sensitive to moisture. Well-dried samples of both of them, stored at room temperature and in the dark, are oxidized slowly. Their solutions are very air sensitive, and the aqueous ones decompose (even under a nitrogen atmosphere) on prolonged standing. The $K[Cr_2H(CO)_{10}]$ is readily soluble in methanol, ethanol, acetone, and THF, and is less soluble in water. The $K[W_2H(CO)_{10}]$ is moderately soluble in these solvents, and is somewhat more soluble in water. The salts of other cations can be obtained by metathesis in anhydrous acetone, e.g., those of PPN^+ and Bu_4N^+ (see Procedure). These latter are more stable in the solid state than the potassium ones, and are soluble in acetone, THF, CH_2Cl_2, $CHCl_3$, and diethyl ether, moderately soluble in ethanol, and insoluble in water. The $PPN[Cr_2H(CO)_{10}]$ melts at 139–140°, and the $PPN[W_2H(CO)_{10}]$ at 119–120°; the $Bu_4N[Cr_2H(CO)_{10}]$ melts at 70–71°, and the $Bu_4N[W_2H(CO)_{10}]$ at 62–63°.†

The solutions of the compounds $K[Cr_2H(CO)_{10}]$ and $K[W_2H(CO)_{10}]$, as well as of the PPN^+ and Bu_4N^+ salts in THF, exhibit three characteristic ν_{CO} infrared

*Prepared as described in the literature.[9]

†Melting points are determined in sealed, evacuated capillaries.

bands: those for the former are at 2029(w), 1940(vs), and 1879(m) cm^{-1}, and those for the latter are at 2042(w), 1938(vs), and 1879(m) cm^{-1}. The ^1H nmr in $(CD_3)_2CO$ against tetramethylsilane, shows δ-19.10 ppm and δ-12.42 ppm, respectively. The $[Cr_2(CO)_{10}(\mu\text{-H})]^-$ anion can interact with donor molecules to give adducts through its electron-deficient Cr-H-Cr system,[4] and with HCl in methanol to give derivatives of the $[Cr(CO)_5Cl]^-$ species.[10] On reaction with HgX_2 (X = Cl, Br, I, or NCS) or I_2, it is rapidly converted into the $[Cr_2(CO)_{10}(\mu\text{-}X)]^-$ or $[Cr(CO)_5X)^-$ species; but if allowed to react in the presence of L or L-L ligands, it yields $LCr(CO)_5$ or $(L\text{-}L)Cr(CO)_4$ complexes.[5]

References

1. (a) H. Behrens and W. Haag, *Chem. Ber.*, **94,** 312 (1961), and references therein. (b) H. Behrens and J. Vogl, *Chem. Ber.*, **96,** 2220 (1963).
2. U. Anders and W. A. G. Graham, *Chem. Comm.*, 499 (1965).
3. R. G. Hayter, *J. Am. Chem. Soc.*, **88,** 4376 (1966).
4. M. D. Grillone and B. B. Kedzia, *J. Organomet. Chem.*, **140,** 161 (1977).
5. M. D. Grillone, *Transition Met. Chem.*, **6,** 93 (1981).
6. J. L. Petersen, R. K. Brown, and J. M. Williams, *Inorg. Chem.*, **20,** 158 (1981), and references therein.
7. J. Roziere, J. M. Williams, R. P. Stewart, J. L. Petersen, and L. F. Dahl, *J. Am. Chem. Soc.*, **99,** 4497 (1977).
8. (a) G. P. Boldrini and A. Umani-Ronchi, *Synthesis*, 596 (1976). (b) D. J. Darensbourg and M. J. Incorvia, *Inorg. Chem.*, **18,** 18 (1979).
9. J. K. Ruff and W. J. Schlientz, *Inorg. Synth.*, **15,** 84 (1974).
10. M. D. Grillone and B. B. Kedzia, *Transition Met. Chem.*, **4,** 256 (1979).

8. BIS[HALOTETRACARBONYLMANGANESE(I)]

Submitted by FAUSTO CALDERAZZO,* RINALDO POLI,* and DARIO VITALI*
Checked by C.-N. CHAU and A. WOJCICKI†

These compounds were originally prepared by Abel and Wilkinson[1] by way of thermal dissociation of coordinated carbon monoxide from the monomeric halo pentacarbonyl derivatives of manganese(I), according to the following stoichiometry:

$$2\,[MnX(CO)_5] \longrightarrow [Mn_2X_2(CO)_8] + 2\,CO \qquad (1)$$

(X = Cl,Br,I)

*Institute of General Chemistry, University of Pisa, 56100 Pisa, Italy.
†Department of Chemistry, The Ohio State University, Columbus, OH 43210.

For reaction (1), the authors used an inert solvent, namely a mixture of saturated hydrocarbons boiling in the 100–120° range. The reported yields are about 40% due to partial decompositon.

The bromo- and iodo-derivatives, which are more frequently used due to their higher thermal stability with respect to the chloro analog, can be conveniently prepared in a poorly coordinating solvent such as diisopropyl ether, which boils at 68–69°. Presumably, the ether solvent stabilizes the tetracarbonyl intermediate arising from CO dissociation before it dimerizes to the final product. The stabilizing effect of the ether solvent, the low temperature, and the low partial pressure of carbon monoxide during the experiment (due to the elevated vapor pressure of the solvent) all contribute favorably to minimizing the formation of the undesired decomposition products $Mn_2(CO)_{10}$ and MnX_2, and to give excellent yields of the dimeric halo-carbonyl complexes.

$$[MnX(CO)_5] + R_2O \rightleftharpoons [MnX(CO)_4(R_2O)] + CO \qquad (2)$$

$$2\,[MnX(CO)_4(R_2O)] \rightleftharpoons [Mn_2X_2(CO)_8] + 2\,R_2O \qquad (3)$$

Procedure

■ **Caution.** *Carbon monoxide and carbonyl compounds are extremely toxic and all reactions must be carried out in an efficient fume hood. Diisopropyl ether is very peroxidizable and must be handled with extreme care. Excess solvent should be discarded as soon as possible.*

A 100-mL reaction flask equipped with an inert gas inlet and connected to another flask of the same volume through a fine, sintered-glass filter is used.* Under an atmosphere of purified dinitrogen, 2.91 g of $[MnBr(CO)_5]$ (10.59 mmoles) and 50 mL of i-Pr_2O freshly distilled over $Li[AlH_4]$ are introduced into the reaction flask. The suspension is refluxed (the walls of the filter (or condenser) are used as a refluxing chamber) under an initially slightly reduced pressure, and the reaction flask is then connected periodically and briefly to a vacuum line, while the temperature of the external bath is at about 80°. The reaction is monitored by infrared spectroscopy in the carbonyl stretching region, and, after two to three hours, no $[MnBr(CO)_5]$ is observed in solution, while the slightly-soluble, brick-red $[Mn_2Br_2(CO)_8]$ is present as a precipitate. The dimeric bromo-derivative is filtered at room temperature, washed with water (to remove $MnBr_2$), and then with i-Pr_2O, and finally dried under vacuum at room temperature for several hours (yield: 2.2 g; 84.1%),

The product obtained by this procedure is analytically and spectroscopically pure according to the published spectrum[2] and does not require further purification

*The checkers report that a water-cooled condenser may be used in place of the sintered-glass filter.

for most chemical purposes. Operating under similar conditions, except for the longer reaction time (about 6 hr), a 90% yield of $[Mn_2I_2(CO)_8]$ is similarly obtained. The longer reaction time required is in agreement with the observed[3] slower kinetics for CO substitution from $[MnI(CO)_5]$ compared to $[MnBr(CO)_5]$.

It was found that for both the bromo- and the iodo-derivatives, prolonged refluxing in *i*-Pr_2O leads to decomposition to $[Mn_2(CO)_{10}]$ and MnX_2.

References

1. E. W. Abel and G. Wilkinson, *J. Chem. Soc.*, 1501 (1959).
2. M. A. El-Sayed and H. D. Kaesz, *Inorg. Chem.*, **2,** 158 (1963).
3. A. Wojcicki and F. Basolo, *J. Am. Chem. Soc.*, **83,** 525 (1961).

9. SODIUM HEXACARBONYLNIOBATE(-I)

Submitted by FAUSTO CALDERAZZO* and GUIDO PAMPALONI*
Checked by J. E. ELLIS†

$$2\ NbCl_5 + 12\ CO + 6\ Mg \xrightarrow{\text{pyridine}} Mg[Nb(CO)_6]_2 + 5\ MgCl_2$$

$$Mg[Nb(CO)_6]_2 + 2\ NaOH \longrightarrow 2\ Na[Nb(CO)_6] + Mg(OH)_2$$

$$Na[Nb(CO)_6] \xrightarrow{\text{tetrahydrofuran}} Na[Nb(CO)_6]\cdot THF$$

The literature covering the preparative methods of the hexacarbonylniobate(-I) anion has been reviewed.[1] All the known methods require the reduction of $NbCl_5$ by sodium[2] or by Na-K alloy[1] under a high pressure of carbon monoxide, but only a 13.9% yield of $[K(CH_3OCH_2CH_2)_2O)_3][Nb(CO)_6]$ has been achieved.

By using pyridine as the solvent and magnesium/zinc as a reducing agent, $NbCl_5$ can be carbonylated to the $[Nb(CO)_6]^-$ anion at room temperature and at *atmospheric pressure* of carbon monoxide. The method does not require the use of the hazardous alkali metals and gives better yields than the ones reported earlier.

Procedure

■ **Caution.** *All operations must be carried out in an efficient fume hood, due to the poisonous nature of carbon monoxide. Use care when handling* CaH_2 *and avoid flame.*

*Institute of General Chemistry, University of Pisa, 56100 Pisa, Italy.
†Department of Chemistry, University of Minnesota, Minneapolis, MN 55455.

Sodium Hexacarbonylniobate(-I) Tetrahydrofuran

Purification of the pyridine solvent is critical, and must be carried out with great care, to eliminate water. It is first dried over KOH pellets, refluxed over CaH_2 under prepurified nitrogen for about 36 hours and then distilled from it. The distillation heads must be tested for the presence of water by addition of sodium and naphthalene, and are discarded until a persistent violet color is observed. The solvent to be used in the reaction (1100 mL) is then directly distilled into the reaction flask containing the magnetically stirred mixture of magnesium (12 g, 0.49 mole) and zinc (20 g, 0.31 mole) powders.* At the end of the distillation, the flask is evacuated and is then filled with carbon monoxide.† In a stream of carbon monoxide, freshly sublimed‡ niobium pentachloride (18.26 g, 67.6 mmoles) is added. The reaction flask is then connected to a gas buret containing carbon monoxide over mercury, and the reaction mixture is vigorously stirred magnetically. The temperature is maintained at 16° by an external water bath operated by a thermostat.

Reduction of $NbCl_5$ occurs rapidly, as denoted by the change in color of the reaction mixture from yellow-orange to violet and then blue. When the reaction mixture is greenish, a fast carbon monoxide absorption starts, and continues at an approximate rate of 2 L/hr up to a molar CO/Nb ratio of about three.§ After that, a considerable decrease in the gas absorption rate is noted. The reaction mixture is vigorously stirred for about 30 hours when the molar CO/Nb ratio reaches approximately six. The red-brown suspension is decanted after about 12 hours, and then filtered under an atmosphere of carbon monoxide.**

The filtered solution is evaporated under vacuum to dryness†† at room tem-

*Magnesium and zinc powders are reagent grade and are used without further purification. Activation of magnesium by iodine can be used, but is unnecessary.

†The submitters used either pure, dry carbon monoxide, free from molecular hydrogen, or commercially available carbon monoxide containing 1–3% of molecular hydrogen, with equally satisfactory results.

‡Commercially available $NbCl_5$ must be purified by double sublimation at 100–110°/~10^{-2} torr. In the case of particularly impure samples, the chloride must be treated with refluxing sulfinyl chloride,[3] followed by evaporation of the solvent and double sublimation of the resulting solid residue.

§The checker did not measure CO absorption and maintained the CO pressure at approximately 800 torr throughout the reaction period (~30 hr). C.P. grade CO was purified of CO_2 by passing through a column of Ascarite, molecular sieves, and activated copper.

**Gasvolumetric measurements of the filtered solution by decomposition with iodine showed that the carbonylation yields vary 44–60%, depending on the operating conditions. Atomic absorption analyses of the cationic metals contained in the solution give magnesium/zinc molar ratios ranging from 9 to 24, showing that magnesium is mainly responsible for the reduction. However, the simultaneous presence of both metals gives the best results. The checker used a medium porosity filtration unit and filtered under an atmosphere of nitrogen.

††From now on, the operations are carried out under reduced pressure, as indicated, or under an atmosphere of prepurified argon.

perature ($17°/10^{-1}$ torr), and the resulting deep brown residue is treated rapidly with 500 mL of a 1.2 *M* aqueous solution of NaOH while the temperature is controlled with an external ice bath. The resulting orange mixture is transferred into a separatory funnel equipped with an upper stopcock for operating under an inert atmosphere, and extracted several times with diethyl ether free of peroxides (distilled over $Li[AlH_4]$). The nature of the suspension is such that the precipitate of $Mg(OH)_2$ can easily be eliminated by way of the lower exit of the separatory funnel, together with the aqueous layer.

The combined diethyl ether extracts are evaporated to dryness initially by a water pump and finally by a mechanical pump. The oily yellow residue becomes a microcrystalline powder by addition of 50 mL of prepurified tetrahydrofuran. After cooling the resulting suspension to dry-ice temperature, the crude hexacarbonylniobate(-I) is collected by filtration and then recrystallized by dissolution in 300 mL of tetrahydrofuran at 40°, filtration, and cooling to room temperature. Completion of the crystallization is achieved by cooling the mother liquor to dry-ice temperature overnight. The sodium derivative is collected by filtration and dried under vacuum for about five hours. Yield: 9.4 g, corresponding to 39.1%.* *Anal.* Calcd. for $[Na(C_4H_8O)][Nb(CO)_6]$; $C_{10}H_8NaNbO_7$: CO, 47.2; Nb, 26.1. Found: CO, 48.0; Nb, 25.0

Properties

The sodium derivative, $Na[Nb(CO)_6]$, stabilized by tetrahydrofuran, is a yellow-orange solid that is extremely sensitive to oxygen, and it can contain variable amounts of tetrahydrofuran. Some attempts to reduce the amount of tetrahydrofuran present in the solid (by evacuation at 20°) have resulted in complete decomposition. When a product with a low content of tetrahydrofuran is desired, the treatment of the solid under vacuum should be done carefully. If the solid shows signs of blackening, the operation should be discontinued immediately. The sodium derivative is quite soluble in diethyl ether, sparingly soluble in tetrahydrofuran at room temperature, very slightly soluble in dichloromethane, and substantially insoluble in hydrocarbons. It is quite soluble in water, in which it forms stable, yellow-orange solutions under a carbon monoxide atmosphere, at a pH of 7–8 or higher. At pH values even slightly lower than 7, rapid decomposition occurs with evolution of carbon monoxide and molecular hydrogen. The aqueous solution is characterized by a single CO stretching vibration at 1875 cm^{-1} (CaF_2, 0.01 mm cell), which should be compared with the 1862 cm^{-1} value for $[V(CO)_6]^-$ under similar conditions. The tetrahydrofuran solution has a main band at 1860 cm^{-1} and a shoulder at 1887 cm^{-1}, presumably caused by distortion of the octahedral structure by the countercation.

*Yields as high as 47% of recrystallized product based on initial niobium have been obtained.

From the aqueous solution of $[Nb(CO)_6]^-$, the nickel-phenanthroline derivative $[Ni(phen)_3][Nb(CO)_6]_2$ can be precipitated as a brick-red, microcrystalline solid. From a dichloromethane suspension of $Na[Nb(CO)_6]$, the μ-nitrido-bis(triphenylphosphorus[1+]) derivative $[(PPh_3)_2N][Nb(CO)_6]$, is obtained as a yellow solid, soluble in dichloromethane, after addition of $(PPh_3)_2NCl$ and recrystallization from dichloromethane-diethyl ether. The crystal and molecular structures of ionic $[(PPh_3)_2N][Nb(CO)_6]$, determined by X-ray diffraction methods, show that $[Nb(CO)_6]^-$ has an almost perfect octahedral geometry.

References

1. J. E. Ellis and A. Davison, *Inorg. Synth.*, **16**, 68 (1976).
2. (a) R. P. M. Werner and H. E. Podall, *Chem. Ind.* (*London*) 144 (1961); (b) R. P. M. Werner, A. H. Filbey, and S. A. Manastyrskyi, *Inorg. Chem.*, **3**, 298 (1964).
3. C. Brauer, *Handbook of Preparative Inorganic Chemistry,* Vol. 2, 2nd Ed., Academic Press, New York, 1963, p. 1303.

10. DISUBSTITUTED DERIVATIVES OF HEXACARBONYLCHROMIUM(0)

Submitted by M. J. WOVKULICH and J. D. ATWOOD*
Checked by W. KRONE-SCHMIDT and H. D. KAESZ†

$$Cr(CO)_6 + Et_4NCl \longrightarrow [Et_4N][Cr(CO)_5Cl] + CO$$

$$[Et_4N][Cr(CO)_5Cl] + L \longrightarrow [Et_4N][LCr(CO)_4Cl] + CO$$

$$L = PBu_3,\ P(OMe)_3,\ P(OPh)_3$$

(A) $$[Et_4N][LCr(CO)_4Cl] + L' \longrightarrow trans\text{-}[Cr(CO)_4LL'] + Et_4NCl$$

$$L' = PPh_3,\ P(OPh)_3,\ AsPh_3$$

(B) $$trans\text{-}Cr(CO)_4(AsPh_3)L + L' \longrightarrow trans\text{-}[Cr(CO)_4LL'] + AsPh_3$$

$$L = PPh_3,\ PBu_3,\ P(OPh)_3,\ P(OMe)_3$$

The reported procedure for the preparation of *trans*-$[Cr(CO)_4(PBu_3)L]$ ($L = PPh_3$, $P(OPh)_3$) involves heating a mixture of $[Cr(CO)_5L]$ and PBu_3 at 160°.[1] This method can be applied to the synthesis of other disubstituted derivatives of

*Department of Chemistry, State University of New York at Buffalo, Buffalo, NY 14214.
†Department of Chemistry and Biochemistry, University of California, Los Angeles, Los Angeles, CA 90024.

hexacarbonylchromium; however, poor yields or complex product mixtures are obtained. A more attractive synthetic route involves the reaction between the substituted salt $[Et_4N][LCr(CO)_4Cl]$ and the ligand L′ to give $[Cr(CO)_4LL']$. An alternative route exploits the facile displacement of $AsPh_3$ by other ligands from the readily prepared *trans*-$[Cr(CO)_4(AsPh_3)L]$.

General Procedure

■ **Caution.** *All preparations must be carried out in a fume hood due to the evolution of carbon monoxide in these reactions.*

All manipulations are conveniently accomplished using a standard Schlenk apparatus. Tetrahydrofuran (THF) is refluxed over sodium/benzophenone until dry, then distilled under nitrogen prior to use. Dichloromethane is stirred over KOH, then distilled from P_4O_{10}. Ethanol is degassed using freeze-pump-thaw cycles, and is then stored over molecular sieves.

The salt $[Et_4N][Cr(CO)_5Cl]$ is prepared by modification of a procedure reported by Abel and co-workers.[2] A mixture of 4.0 g of $[Cr(CO)_6]$ (1.8×10^{-2} mole) and 3.4 g $Et_4NCl{\cdot}H_2O$ (1.8×10^{-2} mole) in 1,1′-oxybis[2-methoxyethane]/THF (diglyme/THF) (20 mL/20mL) is refluxed under argon for four hours, giving an orange solution. The presence of THF is necessary to eliminate the problem of sublimation of unreacted $Cr(CO)_6$, and it also minimizes decomposition of the product. After the mixture has cooled to room temperature, removal of THF under vacuum is followed by addition of 100 mL of pentane, which causes precipitation of the product, $[Et_4N][Cr(CO)_5Cl]$. This yellow solid is washed with pentane, then dried under vacuum overnight to remove unreacted $[Cr(CO)_6]$.

The substituted salts, $[Et_4N][LCr(CO)_4Cl]$, are prepared by applying an adaptation of the procedure reported by Schenk.[3] To a THF solution (20 mL) of 1.0 g $[Et_4N][Cr(CO)_5Cl]$ (2.8×10^{-3} mole), an excess of the ligand L is added and allowed to react at room temperature under argon for a specified time. The amounts of L used, reaction times, and yields are 2 mL PBu_3 (8.0×10^{-3} mole), 10 minutes, and 59%; 1 mL $P(OMe)_3$ (8.5×10^{-3} mole), 30 minutes, and 72%; and 2 mL $P(OPh)_3$ (7.6×10^{-3} mole), 90 minutes, and 55%.* For L = PBu_3 and $P(OMe)_3$ the reactions are accompanied by noticeable effervescence upon addition of the ligand to the solution. The reactions are stopped by the addition of 100 mL of pentane, which causes the formation of a yellow precipitate. This solid, $[Et_4N][LCr(CO)_4Cl]$, is washed several times with 20-mL portions of pentane to remove all traces of excess ligand, then dried under vacuum.

*The reaction time is very important for obtaining the reported yield.

TABLE I Analyses

Compound	Calculated					Found				
	Cr	P	As	C	H	Cr	P	As	C	H
trans-$[Cr(CO)_4(P(OPh)_3)(PPh_3)]$	7.06	8.41		65.22	4.11	7.00	8.31		65.13	4.20
trans-$[Cr(CO)_4(P(OPh)_3)(AsPh_3)]$	6.66	3.97	9.60	61.55	3.87	6.43	4.27	9.60	61.82	4.09
trans-$[Cr(CO)_4(P(OMe)_3)(PPh_3)]$	9.45	11.26		54.56	4.40	9.49	11.28		54.54	4.43
trans-$[Cr(CO)_4(P(OMe)_3)(AsPh_3)]$	8.75	5.21	12.61	50.52	4.07	8.76	5.59	12.22	50.38	4.32
trans-$[Cr(CO)_4(P(OMe)_3)(P(OPh)_3)]$	8.69	10.35		50.18	4.04	8.64	10.56		50.40	4.35
trans-$[Cr(CO)_4(PBu_3)(PPh_3)]$	8.27	9.85		64.96	6.73	8.02	9.91		65.08	6.91
trans-$[Cr(CO)_4(PBu_3)(AsPh_3)]$	7.73	4.61	11.14	60.72	6.29	7.90	4.91	11.64	60.65	6.42
trans-$[Cr(CO)_4(PBu_3)(P(OPh)_3)]$	7.66	9.42		60.17	6.24	7.48	9.37		60.40	6.33

Procedure A

The neutral complexes, *trans*-[$Cr(CO)_4LL'$] (L,L′ = PPh_3, PBu_3, $P(OPh)_3$, $P(OMe)_3$, $AsPh_3$), are prepared by the addition of 10 mL of a CH_2Cl_2 solution of L′ to a Schlenk flask containing 1.0 g of [Et_4N][$LCr(CO)_4Cl$] ($\sim 1.9 \times 10^{-3}$ mole) and 10 mL of deoxygenated ethanol. The amounts of L′ used are 0.5 g of PPh_3 (1.9×10^{-3} mole), 0.5 mL of $P(OPh)_3$ (1.9×10^{-3} mole), and 0.6 g of $AsPh_3$ (1.9×10^{-3} mole). After being stirred for one hour under argon at room temperature, the originally orange solution turns green-yellow. Solvent removal gives a green-yellow solid which is extracted with hexane. This hexane solution, containing *trans*-[$Cr(CO)_4LL'$], is filtered, and the solvent is pumped off, giving a yellow or white solid, which is recrystallized from C_2H_5OH/CH_2Cl_2. The crystals are washed with pentane, then dried under vacuum.

Procedure B

An alternate method for the preparation of *trans*-[$Cr(CO)_4LL'$] involves the facile displacement of $AsPh_3$ by L′ from *trans*-[$Cr(CO)_4(AsPh_3)L$] (L,L′ = PPh_3, PBu_3, $P(OPh)_3$, $P(OMe)_3$). Typically, *trans*-[$Cr(CO)_4(AsPh_3)L$] and a slight excess of L′ are stirred in THF under argon for 18–24 hours at room temperature, although moderate heating (40°–50°) may be applied to hasten the reaction. The orange solution turns green-yellow, and solvent removal gives a green-yellow solid. This solid is recrystallized from C_2H_5OH/CH_2Cl_2 to give the pure product *trans*-[$Cr(CO)_4LL'$]. It was found that this procedure is the best route to *trans*-[$Cr(CO)_4(PBu_3)(P(OMe)_3)$], which is prepared from *trans*-[$Cr(CO)_4(PBu_3)(AsPh_3)$] and $P(OMe)_3$. Analyses are given in Table I.

TABLE II Physical Properties

Compound	Color	Melting point	Yield[a]
trans-[$Cr(CO)_4(P(OPh)_3)(PPh_3)$]	Yellow	149–150°	30%
trans-[$Cr(CO)_4(P(OPh)_3)(AsPh_3)$]	Yellow	138–139°	20%
trans-[$Cr(CO)_4(P(OMe)_3)(PPh_3)$]	Yellow	132–133°	40%
trans-[$Cr(CO)_4(P(OMe)_3)(AsPh_3)$]	Yellow	119–120°	40%
trans-[$Cr(CO)_4(P(OMe)_3)(P(OPh)_3)$]	White	65–66°	40%
trans-[$Cr(CO)_4(PBu)_3(PPh_3)$]	Yellow	141–142°[b]	30%
trans-[$Cr(CO)_4(PBu_3)(AsPh_3)$]	Yellow	129–130°	29%
trans-[$Cr(CO)_4(PBu_3)(P(OPh)_3)$]	White	47–48°[b]	40%
trans-[$Cr(CO)_4(PBu_3)(P(OMe)_3)$]	Pale yellow oil	—	20%

[a]Reaction times must be observed strictly to achieve the quoted yields. The percentages quoted ar the overall yields from $Cr(CO)_6$. Each compound except [$Cr(CO)_4(PBu_3)(P(OMe)_3)$] was prepare by sequence A.

[b]Reference 1.

TABLE III Infrared Carbonyl Stretching Frequencies[a]

Compound	ν(CO), cm^{-1}		
trans-$[Cr(CO)_4(P(OPh)_3)(PPh_3)]$	2023(w)	1958(w)	1914(s)
trans-$[Cr(CO)_4(P(OPh)_3)(AsPh_3)]$	2020(w)	1958(w)	1914(s)
trans-$[Cr(CO)_4(P(OMe)_3)(PPh_3)]$	2017(w)	1952(w)	1903(s)
trans-$[Cr(CO)_4(P(OMe)_3)(AsPh_3)]$			1903(s)
trans-$[Cr(CO)_4(P(OMe)_3)(P(OPh)_3)]$	2038(w)	1970(w)	1925(s)
trans-$[Cr(CO)_4(PBu_3)(PPh_3)]$	2004(w)	1931(w)	1882(s)
trans-$[Cr(CO)_4(PBu_3)(AsPh_3)]$	2003(w)	1930(w)	1882(s)
trans-$[Cr(CO)_4(PBu_3)(P(OPh)_3)]$	2018(w)	1943(w)	1903(s)
trans-$[Cr(CO)_4(PBu_3)(P(OMe)_3)]$	2012(w)	1940(w)	1892(s)

[a]The spectra were recorded on a Beckman 4240 spectrophotometer with hexane as solvent.

Properties

All of the disubstituted complexes listed in Table II are solids at room temperature (with the exception of the oil, *trans*-$[Cr(CO)_4(PBu_3)(P(OMe)_3)]$, are stable to air and light, and are soluble in most organic solvents. Solutions of these compounds are stable in air but exhibit some decomposition when exposed to light. In addition to the physical properties given in Table II, derivatives of $[Cr(CO)_4(PPh_3)L]$ (L = PBu_3, $P(OPh)_3$, $P(OMe)_3$) have been further characterized by way of single-crystal X-ray diffraction studies. The infrared spectra in the carbonyl stretching region are given in Table III.

References

1. S. O. Grim, D. A. Wheatland, and P. R. McAllister, *Inorg. Chem.*, **7**, 161 (1968).
2. E. W. Abel, I. S. Butler, and J. G. Reid, *J. Chem. Soc.*, 2068 (1963).
3. (a) W. A. Schenk, *J. Organomet. Chem.*, **179**, 253 (1979); (b) W. A. Schenk, *J. Organomet. Chem.*, **184**, 195 (1980).

11. PENTACARBONYLRHENIUM HALIDES

Submitted by STEVEN P. SCHMIDT,* WILLIAM C. TROGLER,† and FRED BASOLO*
Checked by MICHAEL A. URBANCIC‡ and JOHN R. SHAPLEY‡

The pentacarbonylrhenium halides, first prepared by Hieber,[1,2] are starting materials for the syntheses of many novel rhenium carbonyl compounds.[3–7] Pho-

*Department of Chemistry, Northwestern University, Evanston, Il 60201.

†Present Address: Department of Chemistry, University of California, San Diego, La Jolla, CA 92093.

‡Chemistry Department, University of Illinois, Urbana, IL 61801.

tochemical,[8,9] vibrational,[10,11] and kinetic[12–15] properties of these molecules have been studied. A rhenium carbonyl halide-alkyl aluminum halide system polymerizes acetylene and is a useful olefin-metathesis catalyst.[16–18]

One member of the halo family, bromopentacarbonylrhenium, is conveniently prepared by direct reaction of an excess of bromine with dirhenium decacarbonyl. Existing literature methods report high yields and facile syntheses.[4,10,19] The chloro analog has been prepared in a similar manner by the action of chlorine on dirhenium decacarbonyl in carbon tetrachloride. Although no explicit mention has been made in the literature of this procedure for the synthesis of $[ReCl(CO)_5]$, it is a straightforward modification of the preparation reported[20] for pentacarbonylchlorotechnetium. This synthesis readily yields the product in acceptable purity and yield.

An alternative procedure involves the photochemical cleavage of the rhenium-rhenium bond in dirhenium decacarbonyl in the presence of a chlorocarbon solvent.[21,22] This method suffers from contamination by the tetracarbonylchlororhenium dimer, which forms in large quantities during photolysis.[21,22] To prevent formation of dimer, the photolysis is performed under an atmosphere of carbon monoxide. The yield of $ReCl(CO)_5$ is essentially quantitative using this procedure.

Existing preparations of the iodo analog involve high temperature[11,23,24] and high pressure.[1,2,25] Photolysis in the presence of iodine under a carbon monoxide atmosphere provides an alternative method of synthesis. Advantages of the photochemical reaction are speed, high yield, and the absence of product contamination by dimer. All reactions may be monitored conveniently by infrared spectroscopy.

Procedure

A. PENTACARBONYLCHLORORHENIUM, $[ReCl(CO)_5]$

$$[Re_2(CO)_{10}] \xrightarrow[CCl_4]{h\nu} 2\ [ReCl(CO)_5]$$

■ **Caution.** *Carbon monoxide is a toxic gas and should be used in an efficient fume hood. Avoid looking directly at the ultraviolet light source. Protective goggles should be worn. Carbon tetrachloride is a carcinogen and must be handled with care in a hood.*

Reagent grade CCl_4 is stirred for two hours over P_4O_{10} (50 mL/0.5 g P_4O_{10}) and then distilled (25 mL) into a 50-mL Schlenk flask (Pyrex), equipped with a Teflon-coated stir bar under an N_2 atmosphere. Dirhenium decacarbonyl (Strem Chemicals—used as received—0.97 g, 1.49 mmoles) is dissolved in the CCl_4.

The solution is then saturated with CO (Matheson, Technical Grade) by bubbling with a syringe needle (through a rubber septum) for 10 minutes.

The photolysis set-up employs a standard 450-W medium-pressure Hanovia mercury arc lamp surrounded by a water-cooled immersion well. The light is directed on the Pyrex Schlenk flask by partially wrapping the well with aluminum foil. The distance from lamp to flask is approximately 3.5 cm. A small magnetic stirrer placed below the flask provides stirring, which is very important to ensure complete reaction and to prevent a buildup of precipitate on the walls of the flask. During photolysis, white $[ReCl(CO)_5]$ crystallizes from solution. Photolysis is terminated when the IR bands due to $[Re_2(CO)_{10}]$ (2073, 2016, and 1976 cm^{-1}) disappear (about 140 min). Aliquots for spectroscopic analysis are withdrawn from the reaction mixture by syringe. A slight positive pressure may accumulate during the reaction.

After solvent is removed under continuous vacuum (0.1 mm) at room temperature, the crude, slightly brownish residue is transferred to a sublimator and sublimed at 80–85° and 0.1 mm. Yield 1.10 g (94.1%, based on $Re_2(CO)_{10}$) [Checkers' yield: 85%]. *Anal.* Calcd. for $[ReCl(CO)_5]$: C, 16.60; Cl, 9.80. Found: C, 16.63; Cl, 9.84.

B. PENTACARBONYLCHLORORHENIUM (ALTERNATIVE PROCEDURE)

$$[Re_2(CO)_{10}] + Cl_2 \xrightarrow{CCl_4} 2\,[ReCl(CO)_5]$$

Reagent-grade CCl_4 is stirred for two hours over P_4O_{10} (50 mL/0.5 g P_4O_{10}), and then distilled (25 mL) into a 50-mL Schlenk flask equipped with a Teflon-coated stir bar under a N_2 atmosphere. The $[Re_2(CO)_{10}]$ (0.32 g, 0.49 mmole) is dissolved in the CCl_4, and the solution is saturated with a stream of chlorine from a syringe needle for one minute. The solution color changes from colorless to yellow-green, and, after two minutes of stirring, a white precipitate forms. The reaction is terminated when the IR bands due to $[Re_2(CO)_{10}]$ disappear (about 140 min).

The solvent and excess chlorine are removed under continuous vacuum (0.1 mm) at room temperature. The white solid is transferred to a sublimator and sublimed at 80–85° and 0.1 mm.* Yield 0.30 g (84.4%, based on $[Re_2(CO)_{10}]$ [Checkers' yield: 75%].

Anal. Calcd. for $[ReCl(CO)_5]$: C, 16.60; Cl, 9.80. Found: C, 16.56; Cl, 9.57.

*Alternatively, the checkers have found that the compound can be purified by dissolving the crude solid in a minimum of warm acetone, adding two volumes of methanol, and recrystallizing at −10°.

C. PENTACARBONYLBROMORHENIUM, $[ReBr(CO)_5]$

$$[Re_2(CO)_{10}] + Br_2 \xrightarrow[\text{hexane}]{} 2\ [ReBr(CO)_5]$$

Hexane (30 mL freshly distilled from sodium benzophenone ketyl) is transferred to a 50-mL Schlenk flask equipped with a Teflon-coated stir bar. The $[Re_2(CO)_{10}]$ (0.89 g, 1.36 mmoles) is added under a stream of N_2, and bromine (0.24 g, 1.50 mmoles) is added to the solution by means of a syringe.* Immediately upon stirring at room temperature, a precipitate forms in the flask. Stirring is continued for 30 minutes, and almost all of the orange bromine color disappears, along with the IR bands of $[Re_2(CO)_{10}]$.

Volatiles are removed under continuous vacuum (0.1 mm) at room temperature. The white powder is transferred to a sublimator and sublimed at 85–90° and 0.2 mm. Yield: 1.04 g (91.3%, based on $[Re_2(CO)_{10}]$).

Anal. Calcd. for $[ReBr(CO)_5]$: C, 14.79; Br, 19.67. Found: C, 14.98; Br, 17.72.

D. PENTACARBONYLIODORHENIUM, $[ReI(CO)_5]$

$$[Re_2(CO)_{10}] + I_2 \xrightarrow[\text{hexane}]{h\nu} 2\ [ReI(CO)_5]$$

■ **Caution.** *See Section A.*

Hexane (30 mL freshly distilled from sodium benzophenone ketyl) is transferred to a 50-mL Pyrex Schlenk flask containing a Teflon-coated stir bar. The $[Re_2(CO)_{10}]$ (0.74 g, 1.13 mmoles) is added under a N_2 stream and dissolved. An excess of iodine (0.435 g, 1.71 mmoles) is then added under a N_2 stream, and the solution is purged with CO for 10 minutes.

The photolysis procedure follows that given in the chloro derivative synthesis (Section 11-A). During irradiation, a white precipitate of $[ReI(CO)_5]$ appears in the violet solution. The reaction is terminated when the IR bands due to $[Re_2(CO)_{10}]$ (2073, 2016, and 1976 cm^{-1}) disappear (about 140 min).†

The solvent is removed under continuous vacuum (0.1 mm) at *room temperature*, and most of the excess I_2 can be pumped off at this point. To remove

*The checkers have found that in dichloromethane, $[Re_2(CO)_{10}]$ reacts instantaneously with bromine. Hence, the carbonyl compound can be "titrated" with bromine until the slightest yellow color persists. This minimizes the amount of residual bromine that must be removed. After solvent removal, the crude solid can be recrystallized by dissolving it in a minimum of warm acetone, adding two volumes of methanol, and cooling to −10°.

†The checkers have found that this reaction was considerably slower than that for the synthesis of the chloro derivative, and thus required a longer period of irradiation (320 min).

TABLE I Carbonyl Stretching Modes (cm^{-1}) (Calibrated against polystyrene)

Compound	In CCl_4			
$[Re(CO)_5Cl]$	n.o.[a]	2049s	2020w	1985m
$[Re(CO)_5Br]$	2154w	2048s	2018w	1987m
$[Re(CO)_5I]$	2149w	2047s	2018w	1991s
$[Re(CO)_4Cl]_2$[26]	2125w	2040s	2005m	1990s
$[Re(CO)_4Br]_2$[26]	2115w	2037s	2000m	1960m
$[Re(CO)_4I]_2$[26]	2111w	2032s	2004m	1968m
$[Re_2(CO)_{10}]$	2073	2016	1976	

[a]n.o. = not observed.

residual amounts of I_2,* the crude solid is transferred to a sublimator and gently heated from 25° to 50° under continuous vacuum with no working coldfinger. When the solid appears nearly white, the temperature is raised to 85°, and the pure $[Re(CO)_5I]$ is sublimed at 0.1 mm. The compound sublimes to a milky-white crystalline solid. Yield: 0.82 g (80.2%) based on $[Re_2(CO)_{10}]$. *Anal.* Calcd. for $[ReI(CO)_5]$: C, 13.25; I, 28.00. Found: C, 13.37; I, 28.18.

Properties

Pentacarbonylrhenium halides are white, crystalline solids that exhibit moderate solubility in nonpolar organic solvents. The iodo derivative is most soluble, and the chloro analog least soluble. All the complexes are stable in air at room temperature; however, heating the solutions or solids results in formation of the tetracarbonylrhenium halide dimers.

As mentioned previously, the infrared spectrum in the carbonyl stretching region of the reaction mixtures is very useful in determining the extent of reaction. The presence of dimeric halide impurities is also easily detected (Table I).

Note. Weak peaks at higher frequency may not appear in solution IR spectra due to limited solubility.

References

1. W. Hieber and H. Schulten, *Z. anorg. allg. Chem.*, **243,** 164 (1939).
2. W. Hieber, R. Schuh, and H. Fuchs, *Z. anorg. allg. Chem.*, **248,** 243 (1941).
3. W. Hieber and W. Opavsky, *Chem. Ber.*, **101,** 2966 (1968).
4. R. J. Angelici and A. E. Kruse, *J. Organomet. Chem.*, **22,** 461 (1970).

*Alternatively, the residual I_2 can be removed by dissolving it in 10 mL of ethanol, and then adding 0.30 g of $Na_2S_2O_3$ in 10 mL of water. The crude product is then filtered, washed with water, and dried under vacuum before being sublimed. If desired, the sublimate can be recrystallized by dissolving it in a minimum of THF, adding two volumes of methanol, and cooling to −10°.

5. R. B. King and R. H. Reimann, *Inorg. Chem.*, **15,** 179 (1976).
6. K. P. Darst and C. M. Lukehart, *J. Organomet. Chem.*, **171,** 65 (1979).
7. L. H. Staal, A. Oskam, and K. Vrieze, *J. Organomet. Chem.*, **170,** 235 (1979).
8. D. R. Tyler and D. P. Petrylak, *Inorg. Chim. Acta,* **53,** L185 (1981).
9. D. M. Allen, A. Cox, T. J. Kemp, Q. Sultana, and R. B. Pitts, *J. Chem. Soc., Dalton Trans.*, 1189 (1976).
10. H. D. Kaesz, R. Bau, D. Hendrickson, and J. M. Smith, *J. Am. Chem. Soc.*, **89,** 2844 (1967).
11. G. Keeling, S. F. A. Kettle, and I. Paul, *J. Chem. Soc., Dalton Trans.*, 3143 (1971).
12. F. Zingales, M. Graziani, F. Faraone, and U. Belluco, *Inorg. Chim. Acta,* **1,** 172 (1967).
13. D. A. Brown and R. T. Sane, *J. Chem. Soc. A.*, 2088 (1971).
14. M. J. Blandamer, J. Burgess, S. J. Cartwright, M. Dupree, *J. Chem. Soc., Dalton Trans.*, 1158 (1976).
15. J. Burgess and A. J. Duffield, *J. Organomet. Chem.*, **177,** 435 (1979).
16. W. S. Greenlee and M. F. Farona, *Inorg. Chem.*, **15,** 2129 (1976).
17. C. D. Tsonis and M. F. Farona, *J. Polym. Sci., Polym. Chem. Ed.*, **17,** 1779 (1979).
18. C. Tsonis and M. F. Farona, *J. Polym. Sci., Polym. Chem. Ed.*, **16,** 185 (1979).
19. E. Horn and M. R. Snow, *Aust. J. Chem.*, **33,** 2369 (1980).
20. J. C. Hileman, D. K. Huggins, and H. D. Kaesz, *Inorg. Chem.*, **1,** 933 (1962).
21. M. Wrighton and D. Bredesen, *J. Organomet. Chem.*, **50,** C35 (1973).
22. M. S. Wrighton and D. S. Ginley, *J. Am. Chem. Soc.*, **97,** 2065 (1975).
23. M. A. Lynch, Jr., W. J. Sesny, and E. O. Brimm, *J. Am. Chem. Soc.*, **76,** 3831 (1954).
24. E. W. Abel, M. M. Bhatti, K. G. Orell, and V. Sik, *J. Organomet. Chem.*, **208,** 195 (1981).
25. L. Vancea and W. A. G. Graham, *J. Organomet. Chem.*, **134,** 219 (1977).
26. R. Colton and J. E. Knapp, *Aust. J. Chem.*, **25,** (1972).

Chapter Two

COMPOUNDS OF BIOLOGICAL INTEREST

12. METAL COMPLEXES OF SACCHARIN

Submitted by S. Z. HAIDER,* K. M. A. MALIK,* and K. J. AHMED*
Checked by GEORGE B. KAUFFMAN and MOHAMMAD KARBASSI†

Saccharin ($C_7H_5SO_3N$), also called 2,3-dihydro-3-oxobenzisosulfonazole; 1,2-benzisothiazoline-3-(2*H*)one 1,1-dioxide; or *o*-benzosulfimide, is widely used as an artificial sweetener. In 1957 Allen and co-workers first produced evidence that saccharin, when implanted into the bladders of mice, might produce urinary bladder carcinomas.[1] Since then, vigorous research activity has been directed toward studying the effect of saccharin on human metabolism.[2,3] Most of these studies were aimed at elucidating its physiological activity. Whether saccharin poses a serious danger to human consumption is still a matter of controversy.[4] Reports on the metal derivatives of saccharin are scanty, and the few that have been reported appear rather confusing.[5–7] The present study is the first systematic attempt to synthesize and characterize the complexes of saccharin with iron(II), cobalt(II), nickel(II), copper(II), and zinc(II), which play interesting roles in human metabolism.

*Department of Chemistry, University of Dacca, Dacca-2, Bangladesh.
†Department of Chemistry, California State University, Fresno, CA 93740.

Bis(*o*-sulfobenzoimido)metal(II) hexahydrates*

[metal(II) = iron(II), cobalt(II), nickel(II), copper(II), and zinc(II)]

Since the nitrogen in the five-membered ring in the saccharin molecule is sp^2 hybridized, the lone pair of electrons on it is assumed to be in a pure *p*-orbital. Hence a substantial amount of conjugation of this *p*-orbital with the *d*-orbitals of sulfur is expected, and this is supported by evidence from crystal structure analysis. The lone pair of electrons is thus unavailable for donation to a metal ion. In the following preparative method, the saccharin anion in its sodium salt is introduced into the medium to facilitate the build-up of the metal-nitrogen linkage.

A. TETRAAQUA-BIS(*o*-SULFOBENZOIMIDO)NICKEL(II) DIHYDRATE

$$Ni(NO_3)_2 \cdot 6H_2O + 2NaNSO_3H_4C_7 \longrightarrow [Ni(C_7H_4SO_3N)_2(H_2O)_4] \cdot 2H_2O + 2NaNO_3$$

Procedure

Nickel nitrate hexahydrate (0.98 g, 0.0034 mole) and 1.6 g of the sodium salt of saccharin (0.0078 mole) are dissolved in 100 mL of water. The solution is heated gently on a water bath until the volume is reduced to 50 mL. It is then allowed to stand overnight, whereupon green crystals separate. The mixture is cooled in an ice bath, and the crystals are collected on a filter paper, washed with two 10-mL portions of cold water, and dried over silica gel. The yield of product obtained is 1.36 g (75.1%). *Anal.* Calcd. for $NiS_2C_{14}N_2H_{20}O_{12}$: Ni, 11.1; S, 12.1; C, 31.6; N, 5.3; H, 3.7. Found: Ni, 11.0; S, 12.1; C, 32.1; N, 5.2; H, 3.8.

This procedure can be used to prepare the following compounds.

B. TETRAAQUA-BIS(*o*-SULFOBENZOIMIDO)COPPER(II) DIHYDRATE

$$CuSO_4 \cdot 5H_2O + 2NaNSO_3H_4C_7 + H_2O \longrightarrow [Cu(C_7H_4SO_3N)_2(H_2O)_4] \cdot 2H_2O + Na_2SO_4$$

Copper sulfate pentahydrate: 0.88 g (0.0034 mole); sodium saccharinate: 1.60 g (0.0078 mole). The yield of greenish-blue, crystalline product is 1.54 g (84.1%).

**o*-Sulfobenzoimido is the 1,2-benzisothiazoline-3-(2*H*)one 1,1-dioxidate anion.

Anal. Calcd. for $CuC_{14}N_2S_2H_{20}O_{12}$: Cu, 11.8; C, 31.3; N, 5.2; S, 11.9; H, 3.7. Found: Cu, 11.7; C, 31.3; N, 5.2; S, 11.8; H, 3.5.

C. TETRAAQUA-BIS(*o*-SULFOBENZOIMIDO)COBALT(II) DIHYDRATE

$$CoCl_2 \cdot 6H_2O + 2NaNSO_3H_4C_7 \longrightarrow [Co(C_7H_4SO_3N)_2(H_2O)_4] \cdot 2H_2O + 2NaCl$$

Cobalt chloride dihydrate: 0.82 g (0.0034 mole); sodium saccharinate: 1.6 g (0.0078 mole). The yield of rose-red, crystalline product is 1.49 g (81.7%).

D. TETRAAQUA-BIS(*o*-SULFOBENZOIMIDO)IRON(II) DIHYDRATE

$$FeSO_4 \cdot 7H_2O + 2NaNSO_3H_4C_7 \longrightarrow [Fe(C_7H_4SO_3N)_2(H_2O)_4] \cdot 2H_2O + Na_2SO_4 + H_2O$$

Iron(II) sulfate heptahydrate: 0.95 g (0.0034 mole); sodium saccharinate: 1.6 g (0.0078 mole). The yield of greenish-yellow, crystalline product is 1.08 g (60.0%).

E. TETRAAQUA-BIS(*o*-SULFOBENZOIMIDO)ZINC(II) DIHYDRATE

$$Zn(NO_3)_2 \cdot 6H_2O + 2NaNSO_3H_4C_7 \longrightarrow [Zn(C_7H_4SO_3N)_2(H_2O)_4] \cdot 2H_2O + 2NaNO_3$$

Zinc nitrate hexahydrate: 1.00 g (0.0034 mole); sodium saccharinate: 1.60 g (0.0078 mole). The yield of white, crystalline product is 0.95 g (51.6%).

Properties

All the metal-saccharin complexes described here exhibit similar properties. They are found to be non-hygroscopic, stable in air, very soluble in pyridine and *N,N*-dimethyl formamide, and slightly soluble in cold water. They give almost identical solid state (Nujol) infrared spectra, characteristic peaks of which are 3500–3100 cm^{-1}, s.b[ν(O—H) of different types of water molecules present in the complex]; 3090 cm^{-1}, s[ν(C—H)]; 1610 cm^{-1}, s[ν(C=O)]; 1570 cm^{-1}, s[ν(C—C), mixed with (H—O—H) deformation of lattice water]; 1280 cm^{-1}, vs[asymmetric stretch of SO_2]; 1260 cm^{-1}, s[δ(C═O)]; 1150 cm^{-1}, vs[symmetric stretch of SO_2]; 1115, 1050, 1000 cm^{-1} [in-plane bending of aromatic C—H]; 755 cm^{-1}, m[out-of-plane bending of aromatic C—H].

Single crystal X-ray analyses of the cobalt, nickel, and copper complexes reveal that they are isostructural and crystallize in the monoclinic space group $P2_1/c$ with a = 7.908 Å, b = 16.136 Å, c = 7.688 Å, β = 99.60° (cobalt-saccharin complex); a = 7.918 Å, b = 16.139 Å, c = 7.651 Å, β = 99.94° (nickel-saccharin complex); a = 8.384 Å, b = 16.327 Å, c = 7.328 Å, β = 101.08° (copper-saccharin complex). All the metal ions have octahedral geometry with four water molecules and the two saccharin anions in the *trans* position. The remaining two water molecules are present as lattice water. The octahedron in tetraaqua-bis(*o*-sulfobenzoimido)copper(II) dihydrate is subject to large tetragonal distortion.[8–10] The metal-nitrogen bond distances in the complexes are: 2.200 Å (cobalt complex), 2.154 Å (nickel complex), and 2.061 Å (copper complex).

The complexes in a DMF medium show the following absorption maxima:

Complexes	Absorption maxima
$Fe(C_7H_4SO_3N)_2 \cdot 6H_2O$	427 nm 285 nm
$[Co(C_7H_4SO_3N)_2(H_2O)_4] \cdot 2H_2O$	525 nm 280 nm
$[Ni(C_7H_4SO_3N)_2(H_2O)_4] \cdot 2H_2O$	680 nm 400 nm 280 nm
$[Cu(C_7H_4SO_3N)_2(H_2O)_4] \cdot 2H_2O$	784 nm 355 nm
$[Zn(C_7H_4SO_3N)_2(H_2O)_4] \cdot 2H_2O$	— —

Thermogravimetric analyses show that the cobalt, nickel, and zinc complexes lose all six water molecules within the temperature range 100–130°. The copper and iron complexes lose four of the water molecules in this temperature range, and the remaining two coordinated water molecules are lost at 230°. The zinc-saccharin complex shows the largest thermal stability. For this compound pyrolysis of the carbonaceous material begins at ~400°. The iron complex decomposes at ~300°, the cobalt complex at ~350°, the nickel complex at ~370°, and the copper complex at ~270°; the end product in each case is the corresponding metal oxide.

References

1. M. J. Allen, E. Bayland, C. E. Dukes, E. S. Horning, and J. G. Watson, *Br. J. Cancer,* **11,** 212 (1957).
2. M. J. Coon, *Proc. Int. Congr. Pharmacol.,* **6,** 117 (1975).
3. I. C. Munro, C. A. Modie, D. Krewski, and H. C. Grice, *Toxicol. Appl. Pharmacol.,* **32** (3), 513 (1975).

4. P. M. Priebe and G. B. Kauffman, *Minerva,* **18,** 556 (1980).
5. A. Tyabji and C. Gibson, *J. Chem. Soc.*, 450 (1952).
6. W. U. Malik and C. L. Sharma, *Indian J. Chem.*, **7,** 920 (1969).
7. H. G. Biedermann, G. Rossmann, and K. E. Schwarzhan, *Z. Naturforsch. B.*, **26** (5), 480 (1971).
8. K. J. Ahmed, S. Z. Haider, K. M. A. Malik, and M. B. Hursthouse, *Inorg. Chim. Acta,* **56,** L37 (1981).
9. S. Z. Haider, K. M. A. Malik, K. J. Ahmed, H. Hess, H. Riffel, and M. B. Hursthouse, *Inorganic Chemica Acta,* **72,** 21 (1983).
10. S. Z. Haider and K. M. A. Malik, *J. Bangladesh Acad. Sci.*, **6,** 119 (1982).

13. *cis-* AND *trans*-DICHLORO-BIS(NUCLEOSIDE)PALLADIUM(II), *cis-* AND *trans*-BIS(NUCLEOSIDATO)PALLADIUM(II), AND *cis-* AND *trans*-BIS(NUCLEOSIDE)BIS(NUCLEOSIDE′)PALLADIUM(II) DICHLORIDE OR TETRAKIS(NUCLEOSIDE)PALLADIUM(II) DICHLORIDE

Submitted by NICK HADJILIADIS*
Checked by P. K. MASCHARAK and S. J. LIPPARD†

$$PdCl_2 + 2\ HCl \longrightarrow H_2[PdCl_4]$$

$$H_2[PdCl_4] \text{ or } K_2[PdCl_4] + 2\ \text{nucl} \xrightarrow{0.5\ N\ HCl} cis\text{-}[Pd(nucl)_2Cl_2] + 2\ HCl \text{ or } 2\ KCl$$

$$cis\text{-}[Pd(nucl)_2Cl_2] \underset{1\ N\ HCl}{\overset{pH \sim 6.5}{\rightleftarrows}} cis\text{-}[Pd(nucl\text{-}H^+)_2] + 2\ HCl$$

$$cis\text{-}[Pd(nucl)_2Cl_2] + 2\ \text{nucl}' \xrightarrow{H_2O} cis\text{-}[Pd(nucl)_2(nucl')_2]Cl_2$$

and

$$K_2[PdCl_4] + 2\ \text{nucl} \xrightarrow{pH \sim 6.5} trans\text{-}[Pd(nucl\text{-}H^+)_2] + 2\ KCl + 2\ HCl$$

$$trans\text{-}[Pd(nucl\text{-}H^+)_2] \underset{pH \sim 6.5}{\overset{1\ N\ HCl}{\rightleftarrows}} trans\text{-}[Pd(nucl)_2Cl_2]$$

$$trans\text{-}[Pd(nucl)_2Cl_2] + 2\ \text{nucl}' \longrightarrow trans\text{-}[Pd(nucl)_2(nucl')_2]Cl_2$$

$$cis\text{- or } trans\text{-}[Pd(nucl)_2Cl_2] + 2\ \text{nucl} \longrightarrow [Pd(nucl)_4]Cl_2$$

Purine nucleosides react very rapidly with Pd(II) in aqueous solutions, producing insoluble precipitates, in many cases in quantitative yields. Several such reactions

*University of Ioannina, Inorganic Chemistry Laboratory, Domboli 31, Ioannina, Greece.
†Department of Chemistry, Massachusetts Institute of Technology, Cambridge, MA 02139.

have been studied[1–7] and have served as models for better understanding the reactions of the antitumor Pt(II) drugs with DNA constituents. Both metals form square planar complexes with similar properties, but Pd(II) aquates more easily and undergoes ligand exchange reactions 10^5 times more rapidly than Pt(II).

Procedure

A. *cis*-[DICHLOROBIS(NUCLEOSIDE)PALLADIUM(II)]

Two mmoles (0.3546 g) of palladium(II) chloride and about 15 mL of 0.5 *N* HCl are placed in a 100-mL beaker equipped with a magnetic stirring bar. The suspension is stirred slowly and heated to boiling, until all of the palladium(II) chloride dissolves. The heating is then stopped and the solution is allowed to cool to room temperature. After cooling, the solution may be filtered, if necessary.

Four mmoles of the proper nucleoside (1.0728 g of inosine or 1.1332 g of guanosine) is dissolved by stirring in about 15 mL of 0.5 *N* HCl in a 50-mL beaker. The two solutions are mixed in a 100-mL beaker and stirred for about two hours at room temperature. The yellow precipitate formed during this time is removed by suction on a medium glass filter (60 mL), and washed with small quantities (10–20 mL) of acetone to remove water. By washing the precipitate with small quantities of diethyl ether (20–30 mL) in small portions using suction on a glass filter, a first drying is achieved. The drying is continued by placing the glass filter containing the product in a desiccator under vacuum at room temperature, in the presence of a drying agent (e.g., KOH, NaOH, or P_4O_{10}) for two to three hours. The drying is completed by keeping the product at about 110° under vacuum, in the presence of P_4O_{10}, until constant weight is achieved. The yield is on the order of 80%.

B. *cis*-[BIS(NUCLEOSIDATO)PALLADIUM(II)]

Two mmoles of the corresponding *cis*-[dichlorobis(nucleoside)palladium(II)] (1.4274 g for the inosine and 1.4882 g for the guanosine complex) are placed in a 100-mL beaker containing a magnetic stirring bar. Upon addition of about 50 mL of distilled water (with stirring) the initial complex dissolves, and, almost instantaneously (after 10–15 min for inosine), a new, yellow precipitate is formed. The pH of this water suspension becomes acidic, due to the liberation of HCl. The stirring is continued for about two hours to attain completion of the reaction, while the pH is kept between 6.0 and 6.5 by the dropwise addition of a solution of 0.1 *N* KOH. The final pH value of the suspension should be 6.0–6.5.

The precipitate is then removed by filtration through a medium glass filter (60 mL) by suction and washed with small quantities (10–20 mL) of ethanol to remove water. The washing is completed with small quantities of diethyl ether (20–30 mL), and the precipitate is dried on the glass filter by stirring it with a small spatula. The glass filter containing the product is then placed in a desiccator and dried over potassium hydroxide (under vacuum) at room temperature for two to three hours. The precipitate is transferred to a small bottle and dried at 110° under vacuum, in the presence of P_4O_{10}, until the weight is constant. The yields are quantitative.

C. *trans*-[BIS(NUCLEOSIDATO)PALLADIUM(II)]

Two mmoles of potassium tetrachloropalladate(II) in a 50-mL beaker is dissolved by stirring in 20 mL of distilled water. Four mmoles of the proper nucleoside (1.0728 g of inosine and 1.1332 g of guanosine) is placed in a 200-mL beaker with 80 mL of distilled water and stirred at room temperature. Inosine dissolves after stirring for a few minutes, whereas guanosine forms a suspension. The two solutions [or the solution of potassium tetrachloropalladate(II) and the guanosine suspension] are mixed in a 200-mL beaker and stirred. A yellow precipitate is formed almost immediately, while the pH of the supernate drops to acidic values. The stirring is continued for about two hours at room temperature while increasing the pH to about 6. This is also the final pH value of the suspension. The precipitate is then removed by suction through a medium glass filter (60 mL) and washed with small portions of ethanol (10–20 mL) (two or three times) to remove water. This is followed by diethyl ether washings (20–30 mL) (in small portions) followed by drying of the precipitate on the glass filter. The drying is continued in a desiccator at room temperature and then at 110° under vacuum, as described above. The yields are quantitative.

D. *trans*-[DICHLOROBIS(NUCLEOSIDE)PALLADIUM(II)]

Two mmoles of the complex *trans*-[bis(nucleosidato)palladium(II)] (1.2816 g for inosine and 1.3420 g for guanosine) and 10 mL of 1 *N* HCl are placed in a 20-mL beaker equipped with a magnetic stirrer and stirred for a few minutes until complete dissolution is achieved. The solution is filtered through a filter paper into a 300-mL beaker, containing an excess (200 mL) of acetone and diethyl ether (1 : 1), and the mixture is stirred with a magnetic stirrer. A yellow precipitate forms immediately. The precipitate is allowed to settle slowly and is removed by filtration through a medium glass filter (60 mL) (by suction). The product is washed with small quantities of acetone and diethyl ether and dried first on the filter glass. The drying is continued in a desiccator, first at room temperature and then at 110° under vacuum, as described above. The yields are quantitative.

E. *cis-* OR *trans-*[BIS(NUCLEOSIDE)BIS(NUCLEOSIDE′)-PALLADIUM(II)] DICHLORIDE AND TETRAKIS(NUCLEOSIDE)PALLADIUM(II) DICHLORIDE

One-half millimole of any of the complexes *cis-* or *trans-*[dichlorobis(nucleoside)palladium(II)] is mixed with 1 mmole of the same nucleoside for the preparation of the [tetrakis(nucleoside)palladium(II)] dichloride complexes, or a different nucleoside, (nucleoside′), for the preparation of the *cis-* or *trans-*[bis(nucleoside)bis(nucleoside′)palladium(II)] dichloride complexes (in the solid state) in a test tube or in a 10-mL conical flask equipped with a small magnetic stirring bar. Then 4 mL of deuterium oxide is added to the mixture with continuous stirring and the container is covered with a cork. When the added nucleosides or (nucleosides′) are inosine or cytidine, the mixture is completely dissolved after stirring for about 5–10 minutes at room temperature. However, in the case of guanosine or xanthosine, the dissolution is achieved only after heating with stirring for about the same time at 50° on a hot plate. The ^1H nmr spectrum of the solution, recorded after the complete dissolution of the mixture in D_2O, is used to indicate the quantitative formation of the corresponding complex as well as the end of the reaction. Distilled water instead of deuterium oxide can be used, if the ^1H nmr spectrum of the final product will not be used prior to isolation. The complex can be isolated by transferring the solution to a 150-mL beaker and adding excess (100 mL) acetone with continuous stirring. Better precipitation is achieved by adding the solution containing the complex dropwise and with stirring to the beaker containing the acetone. The complex then is removed by filtration through a medium glass filter (60 mL), and is washed and dried with small quantities of diethyl ether (20–30 mL), added in small portions. The final drying of the product is achieved as before. The yields are quantitative.

Properties

All of the complexes are yellow powders that decompose without melting. Their decomposition points, elemental analyses, and IR and ^1H nmr spectra have been reported.[4,5] Kurnakoff tests show that the compounds are *cis* and *trans-*$[Pd(nucl)_2Cl_2]$, and that all the ligands are replaced by thiourea in the case of the *cis*, and only the two chlorides in the case of *trans*.[9] The free or palladium-coordinated nucleosides can be observed with ^1H nmr, by mixing the corresponding complexes with excess thiourea in the solid state and dissolving them in D_2O. The geometries of the *cis-* and *trans-*$[Pd(nucl\text{-}H^+)_2]$ complexes are assumed to be the same as the *cis-* and *trans-*$[Pd(nucl)_2Cl_2]$.

The *cis-* and *trans-*$[Pd(nucl\text{-}H^+)_2]$ complexes can be prepared only when a nucleoside containing an easily ionizable imino proton, such as inosine or guanosine, is used. The pK values of the imino proton of these bases fall in the range of 9–9.5 and are reduced by about two logarithmic units upon metal complexation, which usually occurs first at the N_7 position of the purine ring.[4–6,8,10]

References

1. D. J. Nelson, P. L. Yeagle, T. L. Miller, and R. B. Martin, *Bioinorg. Chem.*, **5,** 353 (1976).
2. M. C. Lim and R. B. Martin, *J. Inorg. Nucl. Chem.*, **38,** 1915 (1976).
3. *Ibid.*, **38,** 1911 (1976).
4. G. Pneumatikakis, N. Hadjiliadis, and T. Theophanides, *Inorg. Chim. Acta,* **22,** L_1 (1977).
5. *Ibid.*, *Inorg. Chem.*, **17,** 915 (1978).
6. N. Hadjiliadis and G. Pneumatikakis, *J. Chem. Soc., Dalton,* 1691 (1978).
7. W. M. Beck, J. C. Calabrese, and N. D. Kottmair, *Inorg. Chem.*, **18,** 176 (1979).
8. N. Hadjiliadis and T. Theophanides, *Inorg. Chim. Acta,* **16,** 77 (1976).
9. N. S. Kurnakoff, *J. Prakt. Chem.*, **50,** 483 (1894).
10. N. Hadjiliadis, G. Pneumatikakis, and S. Paraskevas, *Chim. Chron., New Series,* **11,** 11 (1982).

14. [[4,4′4″,4‴-PORPHYRIN-5,10,15,20-TETRAYLTETRAKIS(1-METHYLPYRIDINIUMATO](2-)]-INDIUM(III) PENTAPERCHLORATE

Submitted by P. HAMBRIGHT,* A. ADEYEMO,* A. SHAMIM,* and S. LEMELLE*
Checked by D. K. LAVALLEE, D. MILLER, and A. WHITE†

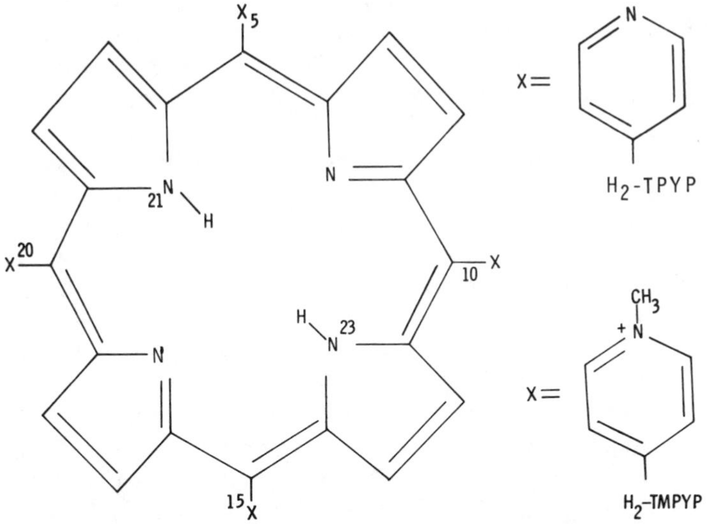

*Department of Chemistry, Howard University, Washington, DC 20059.

†Department of Chemistry, Hunter College of City University of New York, New York, NY 10021.

Metal complexes of water-insoluble porphyrins, such as tetraphenylporphyrin,[1,2] octaethylporphyrin,[3] proto[3] and hematoporphyrin dimethyl esters,[1] and the picket-fence porphyrin[4] have been described in *Inorganic Syntheses*. For many studies, fully water-soluble porphyrins and metalloporphyrins are necessary.[5] The tetra-positive ligand in the title compound, H_2tmpyp, and its metal derivatives have been used extensively where solubility over the full pH range and minimal porphyrin–porphyrin aggregation is desired.[6] A faster, less dangerous, and more convenient synthesis of this ligand than those currently available is presented. The preparation of the indium(III) complex shows the features that are typical for metal-incorporation reactions using water-soluble porphyrins.

A. 5,10,15,20-TETRA(4-PYRIDINYL)PORPHYRIN, H_2-TPYP

The earliest synthesis[7] of H_2-tpyp and later modifications[8,9] involve a 48-hour reaction time, and the use of glass pressure bombs, which can explode. A quicker and safer method uses the Adler[10] condensation conditions as described by Longo,[11] with a simplified work-up method.

Procedure

On a hot plate **under the hood,** 2.5 L of propionic acid in a 6-L Erlenmeyer flask with boiling chips is heated to 90°, followed by the very slow addition of 1 mole of 4-pyridinecarboxaldehyde (95.9 mL) and 1 mole of pyrrole (69.4 mL).

■ **Caution.** *The propionic acid should not be allowed to boil before the addition, because, if the reagents are added too quickly at the boiling point, the resulting exothermic reaction is violent enough to cause the pot to boil over.*

The mixture is heated, allowed to reflux for 45 minutes, and then cooled. The porphyrin is soluble, and instead of evaporating away the solvent,[11] the mixture is poured into a large container, 4 L of water is added, and sufficient sodium acetate trihydrate (about 400 g dissolved in water) is stirred in to bring the pH to about 3.0. The pyridinyl groups are deprotonated at this pH, and, after allowing the porphyrin to settle, the resulting purple precipitate is collected by Büchner filtration (wear gloves), washed with hot N,N-dimethylformamide until the supernate is pale yellow, rinsed with hot water, and dried at 100° Fifteen to eighteen grams of crude H_2-tpyp result. For analytical samples, the H_2-tpyp is dissolved in chloroform and passed through a column of dry alumina (Fisher A-540), and the red porphyrin band is eluted from the slow-moving brown impurities with $CHCl_3$. The H_2-tpyp is then recrystallized from $CHCl_3$/

CH_3OH and dried under vacuum at 130°. *Anal.* Calcd. for $C_{40}H_{26}N_8 \cdot H_2O$: C, 75.46; H, 4.43; N, 17.60. Found: C, 75.55; H, 4.42; N, 17.54.

Properties

The solubility of H_2-tpyp in 17 solvents and the extinction coefficients in $CHCl_3$, pyridine, acetic acid, water (pH 2.3), and 1.0 *M* HCl [443 nm (ϵ = 332,000), 590 nm (ϵ = 14,300) and 639 nm (ϵ = 18,200)] have been tabulated.[12] Metal complexes have been prepared in acetic and formic acids.[7] This porphyrin is relatively insoluble in DMF.

B. 4,4′,4″,4‴-PORPHYRIN-5,10,15,20-TETRAYL-TETRAKIS(1-METHYLPYRIDINIUM) TETRAKIS(4-METHYLBENZENESULFONATE), H_2-TMPYP

For quantities of H_2-tpyp in the milligram range, or for picket fence type pyridinyl porphyrins, which can undergo heat-induced rotational isomerization, N-methylation is best accomplished using excess CH_3I and the porphyrin in $CHCl_3$ at room temperature.[6] For larger amounts of a stable porphyrin, the tosylate method[14] is convenient.

Procedure

Under the hood, in a 1-L three-necked, round-bottomed flask with *overhead* stirring, 15 g of H_2-tpyp is added to 500 mL of N,N-dimethylformamide, followed by 50 mL of methyl (4-methylbenzenesulfonate). The mixture is refluxed for three hours, cooled to room temperature, filtered **(wear gloves),** washed with acetone, dried in air overnight, and finally dried at 130° in a vacuum oven. Yield, 33 g (92%).

■ **Caution.** *Water-soluble porphyrins should not be allowed to touch the skin.*

Anal. Calcd. for $C_{72}H_{66}N_8O_{12} \cdot H_2O$: C, 62.59; H, 4.96; N, 8.10. Found: C, 62.48; H, 5.01; N, 8.13.

Properties

The 4-methylbenzenesulfonate salt of H_2-tmpyp is fully water-soluble and monomeric in aqueous solution. The pK values for protonation of the central nitrogen atoms have been determined.[15] The free-base porphyrin is red in solution at pH 7, and in 1.0 *M* HCl, the green di-acid form has absorption bands at 446 nm,

592 nm, and 643 nm, with molar extinction coefficients of 391,000, 14,000, and 19,000, respectively.

C. THE TITLE COMPOUND

Procedure

One gram of H_2tmpyp 4-methylbenzenesulfonate (0.74 m*M*) and 1 g of $InCl_3$ hydrate (Aldrich, 45.2% In, 3.9 m*M*) are refluxed overnight in 100 mL of 10^{-2} *M* HCl. The solution is then cooled, filtered, and the metalloporphyrin precipitated by the addition of a saturated solution of sodium perchlorate. The complex is isolated by filtration, washed with dilute $NaClO_4$, cold water, and dried under vacuum at room temperature.

■ **Caution.** *Perchlorate complexes are potentially explosive.*

This species can also be precipitated as the iodide or PF_6^- salt,* and the solid brought into solution with Cl^- or NO_3^- anion exchange resins. Yield, 0.8 g (80%). *Anal.* Calcd. for $[In(C_{44}H_{36}N_8)](ClO_4)_5 \cdot 2H_2O$: C, 39.89; H, 3.04; N, 8.46. Found: C, 39.86; H, 2.98; N, 8.32.

Properties

In 10^{-2} *M* HCl, $[In(tmpyp)]^{5+}$ has bands at 424 nm (ϵ = 390,000), 518 nm (ϵ = 3690), 558 nm (ϵ = 23,200), and 597 nm (ϵ = 4660). The spectrum is similar to that of indium(III)-tetraphenylporphyrin.[16] The $[In(tmpyp)]^{5+}$ is fully water soluble, and rapid acid solvolysis occurs only above 3 *M* HCl levels. Refluxing a divalent metal chloride with H_2tmpyp in water, and keeping the pH between 3 and 5, is the general method used to prepare the Cu^{2+}, Zn^{2+}, Ni^{2+}, Mn^{3+}, Fe^{3+}, and Co^{3+} compounds.[6] The 642-nm free-base porphyrin band disappears when the incorporation reaction is complete. The high-pH conditions (a) reduce the concentration of unreactive centrally protonated[15] porphyrins, (b) minimize the extent of acid catalyzed metalloporphyrin solvolysis reactions,[17] and (c) increase the concentration of the often more reactive hydrolyzed metal ion forms.[18]

The tetrakis(3- and 2-pyridinyl)porphyrins,[12] and corresponding N-alkylated derivatives,[19] as well as mixed phenyl/(4-pyridinyl) porphyrins[20] have also been reported.

*This was done by the checkers.

References

1. A. Adler, F. Longo, and V. Varadi, *Inorg. Syn.*, **16,** 213 (1976).
2. E. Johnson and D. Dolphin, *Inorg. Syn.*, **20,** 143 (1980).
3. C. Chang, R. DiNello, and D. Dolphin, *Inorg. Syn.*, **20,** 147 (1980).
4. T. Sorrell, *Inorg. Syn.*, **20,** 161 (1980).
5. P. Hambright, in *Porphyrins and Metalloporphyrins,* K. M. Smith (ed.), Elsevier, Amsterdam, 1975, Chap. 6.
6. P. Hambright and E. Fleischer, *Inorg. Chem.*, **9,** 1757 (1970).
7. E. Fleischer, *Inorg. Chem.*, **1,** 493 (1962).
8. H. Butcher and E. Fleischer, *Inorg. Syn.*, **12,** 256 (1970).
9. K. Ashley, M. Berggren, and M. Cheng, *J. Amer. Chem. Soc.*, **97,** 1422 (1975).
10. A. Adler, F. Longo, J. Finarelli, J. Goldmacher, J. Assour, and L. Korsakoff, *J. Org. Chem.*, **32,** 476 (1967).
11. F. Longo, J. Finarelli, and J. Kim, *J. Hererocyclic Chem.*, **6,** 927 (1969).
12. S. Sugata, S. Yamanouchi, and Y. Matsushima, *Chem. Pharm. Bulletin,* **25,** 884 (1977).
13. N. Datta-Gupta, J. Fanning, and L. Dickens, *J. Coord. Chem.*, **5,** 201 (1976).
14. R. Pasternack, P. Huber, P. Boyd, G. Engasser, L. Francesconi, E. Gibbs, P. Fasella, G. Venturo, and L. deC. Hinds, *J. Amer. Chem. Soc.*, **94,** 4511 (1972).
15. H. Baker, P. Hambright, and L. Wagner, *J. Amer. Chem. Soc.*, **95,** 5942 (1973).
16. M. Bhatti, W. Bhatti, and E. Mast, *Inorg. Nucl. Chem. Letters,* **8,** 133 (1972).
17. A. Adeyemo, A. Valiotti, and P. Hambright, *Inorg. Nucl. Chem. Letters,* **17,** 213 (1981).
18. P. Hambright and P. Chock, *J. Amer. Chem. Soc.*, **96,** 3123 (1974).
19. P. Hambright, T. Gore, and M. Burton, *Inorg. Chem.*, **15,** 2314 (1976).
20. A. Shamim, P. Hambright, and P. Worthington, *J. Chem. Soc. Pakistan,* **3,** 1 (1981).

Chapter Three

STEREOISOMERS

15. OPTICALLY ACTIVE *cis*-UNIDENTATE-DICARBONATO, *cis-cis*-DIUNIDENTATE-CARBONATO, AND UNIDENTATE GLYCINATO COBALT(III) COMPLEXES

Submitted by M. SHIBATA*
Checked by H. G. BRITTAIN†

I. *cis*-UNIDENTATE-DICARBONATO COMPLEXES OF COBALT(III)

The use of a green solution of tricarbonatocobaltate(III) ion[1] enables one to substitute the carbonato ligands successively by other ligands: the first-step substitution gives a dicarbonato complex; the second step, a monocarbonato complex; and the third step, a complex containing no carbonato ligand. Thus, a series of complexes can be synthesized systematically by using the green solution. In particular, the dicarbonato complexes are useful in preparing other complexes.

In the following, preparations of *cis*-diamminedicarbonatocobaltate(III), dicarbonato(1,2-ethanediamine)cobaltate(III), and *cis*-dicarbonatodicyanocobaltate(III) compounds, all starting from the green solution, are given. Besides the preparations, optical resolutions of these compounds are described.

*Faculty of Science, Department of Chemistry, Kanazawa University, Kanazawa 920, Japan.
†Department of Chemistry, Seton Hall University, South Orange, NJ 07079.

A. PREPARATION OF POTASSIUM TRICARBONATOCOBALTATE(III) IN SOLUTION

$$2\ Co(NO_3)_2 + 10\ KHCO_3 + H_2O_2 \longrightarrow 2\ K_3[Co(CO_3)_3] + 4\ CO_2 + 4\ KNO_3 + 6\ H_2O$$

Potassium tricarbonatocobaltate(III) is the starting material. The solid potassium tricarbonatocobaltate(III) trihydrate has been isolated,[2] but it is very unstable. The complex is stable in water containing potassium hydrogen carbonate, and it is therefore used as a solution containing an excess of potassium hydrogen-carbonate.

Procedure

Thirty-five grams (0.35 mole) of potassium hydrogen carbonate is suspended in 20 mL of water, and the slurry is kept in an ice-salt bath. Fifteen grams (0.05 mole) of cobalt(II) nitrate hexahydrate is dissolved in 10 mL of 30% hydrogen peroxide, and the solution is kept in an ice-salt bath. The solution is added dropwise to the ice-cold slurry with stirring.

■ **Caution.** *Add two or three drops at a time. When bubbling stops, again two or three drops are added. This procedure is repeated. Finish adding the whole in about one hour.*

The resulting solution is filtered by suction. The filtrate must be a clear solution and is used immediately as a starting material.

B. POTASSIUM *cis*-DIAMMINEDICARBONATOCOBALTATE(III)

$$[Co(CO_3)_3]^{3-} + 2NH_3 \longrightarrow cis\text{-}[Co(CO_3)_2(NH_3)_2]^{-} + CO_3^{2-}$$

Potassium *cis*-diamminedicarbonatocobaltate(III) monohydrate was originally prepared by the reaction of the green solution of tricarbonatocobaltate(III) with ammonium carbonate.[2] However, the present synthesis is a time-saving modification in that concentrated aqueous ammonia is used in place of the ammonium salt.

Procedure

Seven mL (0.1 mole) of concentrated aqueous ammonia is added to a solution of tricarbonatocobaltate(III) ion (made from 14.5 g of $Co(NO_3)_2\cdot 6H_2O$).

Prepare the green solution of tricarbonatocobaltate(III) carefully. If the green solution is turbid, it must be remade.

The mixture is stirred at room temperature for a few minutes, whereupon the color of the solution becomes inky blue. The resulting solution is cooled rapidly with ice, 50 mL of methanol is added, and the whole is allowed to stand in an ice-bath for a while. The blue precipitate of potassium *cis*-diammine-dicarbonatocobaltate(III) separates out, is collected on a filter, washed with methanol and diethyl ether in turn, and dried at room temperature. The product is recrystallized from a minimum amount of warm water (~35°). The yield is 6 g. *Anal.* Calcd. for $K[Co(CO_3)_2(NH_3)_2]$: C, 9.52; H, 2.38; N, 11.11. Found: C, 9.51; H, 2.36; N, 10.97.

Properties

This complex is unstable in aqueous solution, but stable in water containing potassium hydrogen carbonate. The visible spectrum in solution containing $KHCO_3$ shows ($\tilde{\nu}_{max}$, log ϵ_{max}): (17,400 cm^{-1}, 2.14) and (25,500 cm^{-1}, 2.40).

Isolation of the optically active lithium salt

Attempts to resolve the racemic potassium salt by the usual method of the less soluble diastereomer formation were unsuccessful. In the following procedure, one enantiomer of the lithium salt of the complex is obtained in excess of 50% yield.

Procedure

A solution of 14 g (0.05 mole) of the potassium salt of *cis*-diammine-di(carbonato)cobaltate(III) in 150 mL of cold water is charged on a column containing Dowex 50W-X8 resin in the Li^+ form (100–200 mesh, 7 cm × 15 cm). With the addition of cold water, a blue solution effuses from the column. To the effluent is added 3 g of ammonium carbonate and the solution is concentrated to 50 mL under reduced pressure at 35°. The concentrated solution is filtered and kept in a refrigerator until blue crystals of lithium *cis*-diamminedi(carbonato)cobaltate(III) deposit. This crude product is recrystallized from water, collected on a filter, washed with cold water, ethanol, and diethyl ether, and finally dried under vacuum. The yield is 3 g. *Anal.* Calcd. for $Li[Co(CO_3)_2(NH_3)_2]$: C, 10.92; H, 2.75; N, 12.73. Found: C, 10.92; H, 3.04; N, 12.66. The solubility of this salt in water is considerably lower than that of the potassium salt.

Four grams (0.01 mole) of $(-)_{589}$-bis(1,2-ethanediamine)oxalatocobalt(III) iodide[3] is ground in a mortar with 30 mL of water. To it 1.63 g (0.01 mole) of silver acetate is added, and the mixture is stirred for 10 minutes. The mixture is filtered and the precipitated silver iodide is rinsed with 10 mL of water. A solution of 5 g (0.02 mole) of the lithium salt described above in 50 mL of water is added to the combined filtrate and washing. After 5 g of ammonium carbonate has been added to the mixture, the solution is concentrated to 30 mL under reduced pressure. The concentrated solution is filtered and placed in a refrigerator until blue, powdered crystals separate out. The first crop of crystals is collected by filtration. The filtrate is again concentrated in the presence of ammonium carbonate, and a second crop of crystals is obtained. The combined crystals are washed with 50% cold aqueous methanol, ethanol, and diethyl ether, and finally dried under vacuum. The yield is 3.5 g (~70%). Found: C, 10.83; H, 3.05; N, 12.60.

Properties

This complex is levo rotatory at 589 nm. The CD spectrum in solution shows ($\tilde{\nu}_{max}$, $\Delta\epsilon$): (17,800 cm^{-1}, −2.32), (23,600 cm^{-1}, −0.19) and (26,100 cm^{-1}, +0.26). The complex in solution loses its optical activity with a half-life of ~3 minutes.

C. POTASSIUM DICARBONATO(1,2-ETHANEDIAMINE)COBALTATE(III) MONOHYDRATE

$$[Co(CO_3)_3]^{3-} + en \longrightarrow [Co(CO_3)_2(en)]^- + CO_3^{2-}$$

Potassium dicarbonato(1,2-ethanediamine)cobaltate(III) was originally prepared by the reaction of the green solution with 1,2-ethanediammonium carbonate.[4] The method given here uses free 1,2-ethanediamine instead of its carbonate. This modified method is superior in the time required for the synthesis and in yield.

Procedure

To a clear solution of the tricarbonatocobaltate(III) ion (made from 14.5 g of $Co(NO_3)_2 \cdot 6H_2O$) is added 4.1 mL (0.05 mole) of 98% 1,2-ethanediamine. The mixture is stirred at room temperature for a few minutes, whereupon the color of the solution becomes blue. The solution is cooled rapidly with ice, 50 mL of ethanol is added, and the whole is allowed to stand in an ice-salt bath for a while. An aqueous layer forms and is decanted, and 50 mL of ethanol is added to the residual blue oil. This procedure is repeated until the decanted solution becomes clear. To the final oily material is added 50 mL of ethanol, and the whole is allowed to stand in an ice-salt bath for 30 minutes to crystallize the desired potassium salt. The blue crystals are collected on a filter, washed with methanol and diethyl ether, and finally dried under vacuum at room temperature. The yield is 5 g. *Anal.* Calcd. for $K[Co(CO_3)_2(en)]\cdot H_2O$: C, 16.22; H, 3.37; N, 9.46. Found: C, 16.30; H, 3.43; N, 9.43.

Properties

The visible spectrum in an aqueous solution containing $KHCO_3$ shows ($\tilde{\nu}_{max}$, log ϵ_{max}): (17,500 cm^{-1}, 2.17) and (25,600 cm^{-1}, 2.33).

Resolution

The *cis*-dicarbonato(1,2-ethanediamine)cobaltate(III) ion was first resolved through the $(+)_{589}$-tris(1,2-ethanediamine)cobalt(III) ion.[1] However, the results gave a poor $\Delta\epsilon$ value. A good resolution is accomplished by the use of $(-)_{589}$-bis(1,2-ethanediamine)oxalatocobalt(III).

Procedure

Four grams (0.01 mole) of $(-)_{589}$-bis(1,2-ethanediamine)oxalatocobalt(III) iodide[3] is suspended in 25 mL of water at 35°. To it, 1.63 g (0.01 mole) of silver acetate is added, and the mixture is stirred for five minutes at 35°. The mixture is filtered, and the precipitated silver iodide is washed with 5 mL of warm water. To the filtrate (combined with the washings) 6g (0.02 mole) of potassium dicarbonato(1,2-ethanediamine)cobaltate(III) monohydrate is added, and the resulting solution is cooled in an ice bath, followed by addition of 5 mL of ethanol. The whole is kept in a refrigerator, whereupon a diastereoisomeric salt of the $(-)_{589}$dicarbonato complex separates out as pink-purple crystals. The product is collected on a filter and washed with 10 mL of cold water. It is recrystallized from a minimum amount of water at 10° to increase the optical purity. The recrystallized salt is collected on a filter, washed with cold water, ethanol, and diethyl ether, and finally dried under vacuum. The yield is 2 g. *Anal.* Calcd. for $[Co(C_2O_4)(C_2H_8N_2)_2]\cdot[Co(CO_3)_2(en)]\cdot 2H_2O$: C, 21.44; H, 4.91; N, 18.19. Found: C, 21.33; H, 4.91; N, 18.08.

Properties

The CD spectrum of $(-)_{589}$-$[Co(CO_3)_2(en)]^-$ in water shows ($\tilde{\nu}_{max}$, $\Delta\epsilon$): (17,600 cm^{-1}, +1.91), (23,500 cm^{-1}, +0.22), (26,000 cm^{-1}, −0.10) and (28,800 cm^{-1}, +0.09). (The CD spectrum is measured with a sample converted into the sodium salt by ion-exchange.)

D. TRIS(1,2-ETHANEDIAMINE)COBALT(III) *cis*-DICARBONATODICYANOCOBALTATE(III) DIHYDRATE

$$[Co(CO_3)_3]^{3-} + 2CN^- + [Co(en)_3]^{3+} \longrightarrow [Co(en)_3][Co(CN)_2(CO_3)_2] + CO_3^{2-}$$

It is generally difficult to prepare a cobalt(III) complex containing different kinds of ligands that are considerably separated in the spectrochemical series. The several complexes containing cyanide ion, which occupies the highest position in the series, and ammonia, which is ranked in a middle position, have been prepared by methods to prevent disproportionation reactions.[5] The following procedure gives a mixed cyano complex with carbonate ion.

Procedure

To a freshly prepared green solution of the tricarbonatocobaltate(III), made from 14.5 g of $Co(NO_3)_2 \cdot 6H_2O$, is added 7.7 g (0.12 mole) of potassium cyanide, in small portions. The mixture is stirred at room temperature until it becomes deep red. The red solution is carefully neutralized with a 30% perchloric acid solution and is then filtered. The filtrate is charged on a column of Dowex 1-X8 resin (100–200 mesh, Cl^- form, 5 cm × 10 cm). When the adsorbed band is eluted with a 1 *M* potassium chloride aqueous solution, a red band comes out of the column, followed by a green one. The red effluent is concentrated to a small volume below 30°. After that a small amount of ethanol is added to the concentrate, and the whole is kept in a refrigerator to precipitate the potassium chloride used for the elution. After the removal of this salt, a solution of 12 g (0.025 mole) of tris(1,2-ethanediamine)cobalt(III) bromide[6] in water is added as the precipitant. After the whole is kept in a refrigerator, needle-like, red-brown crystals deposit. These are collected on a filter, washed with cold water, ethanol, and diethyl ether, and finally dried under vacuum. The yield is 1.5 g. *Anal.* Calcd. for $[Co(en)_3] \cdot [Co(CN)_2(CO_3)_2] \cdot 2H_2O$: C, 23.73; H, 5.58; N, 21.60. Found: C, 23.93; H, 5.68; N, 22.10.

Properties

The complex cannot be obtained as a usual salt, such as the potassium or sodium salt, because of its great solubility and because of the instability of the species in aqueous solution. The visible spectrum of the potassium salt, which is obtained by means of ion-exchange techniques, shows ($\tilde{\nu}_{max}$, log ϵ_{max}): (18,300 cm^{-1}, 1.96) and (22,500 cm^{-1}, 2.03).

This complex can be resolved by the $(+)_{589}$-tris(1,2-ethanediamine)cobalt(III) ion, but the diastereoisomer loses its optical activity very quickly.

References

1. M. Shibata, *Nippon Kagaku Zasshi,* **87,** 771 (1966).
2. M. Mori, M. Shibata, E. Kyuno, and T. Adachi, *Bull. Chem. Soc. Japan,* **29,** 883 (1956).
3. F. P. Dwyer, I. K. Reid, and F. L. Garvan, *J. Am. Chem. Soc.,* **83,** 1285 (1961).
4. M. Mori, M. Shibata, E. Kyuno, and K. Hoshiyama, *Bull. Chem. Soc. Japan,* **31,** 291 (1958).
5. (a) H. Siebert, C. Siebert, and K. Wieghardt, *Z. anorg. allg. Chem.,* 380 (1959); (b) K. Konya, H. Nishikawa, and M. Shibata, *Inorg. Chem.,* **7,** 1165 (1968).
6. A. Werner, *Ber.,* **45,** 121 (1912).

II. *cis,cis*-UNIDENTATE-CARBONATO COMPLEXES OF COBALT(III)

The preparation of a complex that derives its dissymmetry from the *cis,cis* distribution of unidentate ligands is generally difficult to achieve. However, the use of a green solution of tricarbonatocobaltate(III) makes such a synthesis of the complex possible. In the following, preparations of *cis,cis*-diamminecarbonatodicyanocobaltate(III), *cis,cis*-diamminedicyanooxalatocobaltate(III), *cis,cis*-diamminecarbonatodinitrocobaltate(III), *cis,cis*-diamminedinitrooxalatocobaltate(III), and *cis,cis*-diamminecarbonatobis(pyridine)cobalt(III) are given. Optical resolutions of these complexes are also described.

A. SODIUM *cis,cis*-DIAMMINECARBONATODICYANOCOBALTATE(III) DIHYDRATE

$$[Co(CO_3)_3]^{3-} + 2CN^- \longrightarrow [Co(CN)_2(CO_3)_2]^{3-} + CO_3^{2-}$$

$$[Co(CN)_2(CO_3)_2]^{3-} + 2NH_3 \longrightarrow \textit{cis,cis}\text{-}[Co(CN)_2(CO_3)(NH_3)_2]^- + CO_3^{2-}$$

Procedure

To a clear green solution of the tricarbonatocobaltate(III) ion (made from 15 g of $Co(NO_3)_2 \cdot 6H_2O$), is added 7.5 g (0.12 mole) of potassium cyanide, and the mixture is stirred at room temperature. A deep red solution of the *cis*-dicarbonatodicyanocobaltate(III) ion is obtained.[2] After the red solution has been neutralized with 30% perchloric acid in an ice bath, 15 g of ammonium perchlorate and 50 mL of concentrated aqueous ammonia are added. The mixture is stirred at 40° for four hours. After filtration, the solution is concentrated to half of its original volume. The concentrated solution is again neutralized and filtered. The filtrate is diluted to triple its initial volume with water and then charged on an ion-exchange column containing Dowex 1-X8 resin in the Cl^- form (5 cm × 20 cm). During elution with a 0.3 *M* sodium chloride solution, an orange band comes out before the others. The effluent containing the orange material is concentrated to 10 mL under reduced pressure. Upon keeping the solution in a refrigerator after the addition of ethanol, yellow-orange crystals of sodium *cis,cis*-diamminecarbonatodicyanocobaltate(III) dihydrate separate. These are recrystallized from warm water (~40°). The crystals are collected on a filter, washed with cold water, ethanol, and diethyl ether, and finally dried under vacuum. The yield is 0.3 g. *Anal.* Calcd. for $Na[Co(CN)_2(CO_3)(NH_3)_2]\cdot 2H_2O$: C, 13.64; H, 3.82; N, 21.22. Found: C, 13.89; H, 4.08; N, 20.84.

Properties

The visible spectrum in water shows ($\tilde{\nu}_{max}$, log ϵ_{max}) (20,700 cm^{-1}, 1.96) and (23,500 cm^{-1}, 2.02).

Resolution Procedure

Two grams (0.005 mole) of $(-)_{589}$-bis(1,2-ethanediamine)oxalatocobalt(III) iodide[3] is suspended in 30 mL of water at 35°, 0.81 g (0.005 mole) of silver acetate is added, and the mixture is stirred for five minutes at 35°. The mixture is then filtered and the precipitated silver halide is washed with 5 mL of warm water (50°). Separately, 2.4 g (0.01 mole) of sodium *cis,cis*-diammine-carbonato-dicyanocobaltate(III) dihydrate is dissolved in 10 mL of water. The solution is added to the filtrate and combined with the washings. After removal of the precipitated material, the mixed solution is kept in a refrigerator, whereupon an orange diastereoisomer of the $(+)_{589}$ form of the carbonato complex precipitates. The product is collected on a filter and washed with cold water. Recrystallization is repeated from water until the CD peak at 24,100 cm^{-1} reaches a constant value. The crystals are collected on a filter, washed with 50% cold, aqueous methanol, ethanol, and diethyl ether, and finally dried under vacuum. The yield is 0.3 g. *Anal.* Calcd. for $[Co(ox)(en)_2][Co(CN)_2(CO_3)(NH_3)_2]\cdot 2H_2O$: C, 21.44; H, 5.16; N, 22.05. Found: C, 21.31; H, 5.25; N, 21.78.

Properties

This diastereoisomeric salt can be broken (cation removal) by passing its solution through a cation-exchange column (Dowex 50W-X8, Na-form). The resulting species retains its optical activity for a day at room temperature. The CD spectrum of $(+)_{589}$-$[Co(CN)_2(CO_3)(NH_3)_2]^-$ shows ($\tilde{\nu}_{max}$, $\Delta\epsilon$): (18,200 cm^{-1}, −0.02), (20,900 cm^{-1}, +0.98), (24,100 cm^{-1}, −1.11) and (29,700 cm^{-1}, +0.25).

B. SODIUM *cis,cis*-DIAMMINEDICYANO-OXALATOCOBALTATE(III) DIHYDRATE

$$cis,cis\text{-}[Co(CN)_2(CO_3)(NH_3)_2]^- + 2H^+ + 2H_2O \longrightarrow cis,cis,cis\text{-}[Co(CN)_2(H_2O)_2(NH_3)_2]^+ + CO_2 + H_2O$$

$$cis,cis,cis\text{-}[Co(CN)_2(H_2O)_2(NH_3)_2]^+ + C_2O_4^{2-} \longrightarrow cis,cis\text{-}[Co(CN)_2(ox)(NH_3)_2]^- + 2H_2O$$

Procedure

To a solution of 13.2 g (0.05 mole) of sodium *cis,cis*-diamminecarbonato-dicyanocobaltate(III) dihydrate in 100 mL of water is added 25 mL of 30% perchloric acid dropwise in an ice-bath until the evolution of carbon dioxide ceases. To the resulting solution is added 18 g (0.10 mole) of potassium oxalate dihydrate, and the mixture is then stirred at 40° for four hours. The solution is filtered after neutralization with a 6 mole/L potassium hydroxide solution, and the filtrate is charged on an ion-exchange column containing Dowex 1-X8 resin in the Cl^- form (100–200 mesh, 4 cm × 30 cm). After the column has been washed with water, the elution is started with a 0.1 mole/L sodium chloride solution. Only one yellow band descends. The effluent is collected and concentrated to 10 mL, and then filtered to remove the deposited sodium chloride. After keeping the filtrate in a refrigerator, yellow crystals deposit. The crude product is recrystallized from the minimum amount of warm water (~50°). The crystals are collected on a filter, washed with cold water, ethanol, and diethyl ether, and finally dried under vacuum. The yield is about 3 g. *Anal.* Calcd. for $Na[Co(CN)_2(ox)(NH_3)_2]\cdot 2H_2O$: C, 16.45; H, 3.45; N, 19.18. Found: C, 16.61; H, 3.37; N, 19.13.

Properties

The visible spectrum in water shows ($\tilde{\nu}_{max}$, log ϵ_{max}): (21,300 cm^{-1}, 1.89) and (24,300 cm^{-1}, 2.01).

Resolution Procedure

Three grams (0.0075 mole) of $(-)_{589}$-bis(1,2-ethanediamine)oxalatocobalt(III) iodide[1] is suspended in 20 mL of water at 50°. To it 0.225 g (0.0075 mole) of silver acetate is added, and the mixture is stirred for five minutes at 50°. The resulting solution is filtered and the precipitated silver halide is washed with 5 mL of warm water at 50°. To the hot filtrate (combined with the washings) is added 4.4 g (0.015 mole) of sodium *cis,cis*-diammine-dicyano-oxalatocobaltate(III) dihydrate. On scratching the sides of the vessel in an ice bath, the orange diastereoisomeric salt of $(-)_{589}$-diamminedicyanooxalatocobaltate(III) separates. The whole is allowed to stand for 0.5 hour in an ice bath. The crude product is collected on a filter and washed with cold water. It is recrystallized repeatedly from warm water (40°), until the $\Delta\epsilon$ value of a main CD peak reaches a constant value (about eight times). The yield is 1 g. *Anal.* Calcd. for $[Co(ox)(en)_2][Co(CN)_2(ox)(NH_3)_2]\cdot 2.5H_2O$: C, 23.62; H, 5.23; N, 20.04. Found: C, 23.37; H, 4.83; N, 20.06.

Properties

The CD spectrum of $(-)_{589}$-diamminedicyanooxalatocobaltate(III) in water shows ($\tilde{\nu}_{max}$, $\Delta\epsilon$): (21,200 cm^{-1}, -1.44), (24,900 cm^{-1}, $+1.24$), (27,800 cm^{-1}, sh), (30,900 cm^{-1}, -0.18), and (34,500 cm^{-1}, sh). (The CD spectrum is measured with a sample converted into the sodium salt by means of the ion-exchange technique mentioned earlier.)

C. POTASSIUM *cis,cis*-DIAMMINECARBONATO-DINITROCOBALTATE(III) HEMIHYDRATE

$$cis\text{-}[Co(CO_3)_2(NH_3)_2]^- + 2H^+ + 2NO_2^- \longrightarrow [Co(NO_2)_2(CO_3)(NH_3)_2]^- + CO_2 + H_2O$$

Procedure

To a solution of 10.1 g (0.04 mole) of potassium *cis*-diammine-dicarbonatocobaltate(III) in 60 mL of water is added 8.5 g (0.10 mole) of potassium nitrite. Then 6 mole/L acetic acid is added to the mixture dropwise with mechanical stirring at room temperature, so that the pH is maintained at 4.5–5.5. The solution is stirred for an additional half-hour. The insoluble, dark green precipitate is removed by filtration, and the filtrate is charged on an ion-exchange column containing 100–200 mesh Dowex 1-X8 resin in the Cl^- form (5 cm × 20 cm). The column is washed with a large amount of ice-cold water, and then the elution of an adsorbed band is carried out with a cold 0.1 mole/L potassium chloride solution. Two bands (colored orange-yellow and orange) descend in this order, and are collected separately. From the first fraction *trans*(NH_3), *cis*(NO_2)-$K[Co(NO_2)_2(CO_3)(NH_3)_2]\cdot H_2O$ can be obtained.[2]

The second fraction is concentrated to 30 mL and filtered. To the filtrate is added 10 mL of ethanol, and the whole is kept in a refrigerator. Powdery, orange crystals of potassium *cis,cis*-diamminecarbonatodinitrocobaltate(III) hemi-hydrate deposit. The material is quickly recrystallized from the minimum amount of warm water (~40°). The crystals are collected on a filter, washed with cold water, ethanol, and diethyl ether, and finally dried under vacuum. The yield is 2 g. *Anal.* Calcd. for *cis,cis*-$K[Co(NO_2)_2(CO_3)(NH_3)_2]\cdot 0.5H_2O$: C, 4.10; H, 2.41; N, 19.11. Found: C, 4.32; H, 2.34; N, 19.28.

Properties

The *cis,cis* complex is unstable in solution at room temperature, and deposits a dark green precipitate. Therefore, it is necessary to perform the recrystallization as quickly as possible. The visible spectrum of the *cis,cis*-isomer in water shows ($\tilde{\nu}_{max}$, log ϵ_{max}): (20,600 cm^{-1}, 2.32). The maximum of the $\pi^* \leftarrow \pi$ band for the *cis,cis*-isomer is at 29,900 cm^{-1} (log ϵ = 3.58).

Resolution Procedure

Three grams (0.0075 mole) of $(-)_{589}$-bis(1,2-ethanediamine)oxalatocobalt(III) iodide[1] is converted into the acetate with 1.3 g (0.008 mole) of silver acetate. To a solution of this resolving agent in 25 mL of warm water, 4.4 g (0.015 mole) of potassium *cis,cis*-diammine-carbonato-dinitrocobaltate(III) hemi-hydrate is added, little by little, with continuous stirring. A red-orange diastereoisomeric salt of the $(-)_{589}$carbonato complex crystallizes on standing in an ice bath. The product is collected on a filter, washed with cold water, and recrystallized three times from water below 35° to avoid decomposition of the carbonato complex. The yield is 0.5 g. *Anal.* Calcd. for $[Co(C_2O_4)(en)_2][Co(NO_2)_2(CO_3)(NH_3)_2]\cdot 2H_2O$: C, 15.34; H, 4.78; N, 20.44. Found: C, 15.45; H, 5.06; N, 20.84. The diastereomer is split as described earlier.

Properties

The CD spectrum of $(-)_{589}[Co(NO_2)_2(CO_3)(NH_3)_2]^-$ in water shows ($\tilde{\nu}_{max}$, $\Delta\epsilon$): (18,800 cm^{-1}, +0.12), (21,700 cm^{-1}, −0.91), and (26,000 cm^{-1}, +0.16).

D. POTASSIUM *cis,cis*-DIAMMINEDINITRO-OXALATOCOBALTATE(III) HEMIHYDRATE

$$[Co(CO_3)_3]^{3-} + (NH_4)_2ox \longrightarrow cis\text{-}[Co(CO_3)(ox)(NH_3)_2]^- + CO_2 + H_2O + CO_3^{2-}$$

$$cis\text{-}[Co(CO_3)(ox)(NH_3)_2]^- + 2H^+ + 2H_2O \longrightarrow cis,cis\text{-}[Co(ox)(H_2O)_2(NH_3)_2]^+ + CO_2 + H_2O$$

$$cis,cis\text{-}[Co(ox)(H_2O)_2(NH_3)_2]^+ + 2NO_2^- \longrightarrow cis,cis\text{-}[Co(NO_2)_2(ox)(NH_3)_2]^- + 2H_2O$$

The reaction of potassium *cis,cis*-diamminedinitrocarbonatocobaltate(III) hemihydrate with oxalic acid gives a *trans*-diammine isomer of this complex. The *cis,cis*-isomer is obtained by the reaction of *cis,cis*-diamminediaquaoxalatocobalt(III) ion with nitrite ion.

Procedure

Twelve grams (0.085 mole) of ammonium oxalate is added to a green solution of tricarbonatocobaltate(III) (made from 14.5 g of $Co(NO_3)_2 \cdot 6H_2O$). The mixture is stirred in a water bath at 60° until the color of the solution becomes blue-violet. The blue-violet solution is filtered, and to the filtrate is added 40 mL of ethanol. The whole is kept in an ice bath for an hour, whereupon a crude product deposits. The crude product is dissolved in 100 mL of water at 50° and filtered. The filtrate is allowed to stand in a refrigerator, whereupon blue-violet crystals of potassium *cis*-diamminecarbonatooxalatocobaltate(III) monohydrate[3] deposit. The crystals are collected on a filter, washed with cold water, ethanol, and diethyl ether, and dried under vacuum. The yield is 7 g.

To a solution of 5.8 g (0.02 mole) of potassium diammine-carbonato-oxalatocobaltate(III) monohydrate in 60 mL of water is added 3 mole/L of nitric acid until the evolution of carbon dioxide ceases. The solution is filtered and the filtrate is allowed to stand in a refrigerator for three hours. Red-violet crystals of *cis,cis*-diamminediaquaoxalatocobalt(III) nitrate[4] deposit. The crystals are collected on a filter and recrystallized by dissolving in water at 40° and adding aqueous nitric acid in the cold. The recrystallized crystals are collected on a filter, washed with 50% cold aqueous methanol, ethanol, and ether, and finally dried in vacuo. The yield is 4 g.

To a solution of 11.2 g (0.04 mole) of *cis,cis*-diammine-diaqua-oxalatocobalt(III) nitrate in 150 mL of water is added 8.5 g (0.1 mole) of potassium nitrite, and then 10 mL of 6 mole/L acetic acid is added to the mixture to adjust the pH to 5. The acidic solution is stirred for an hour at room temperature and then filtered. The filtrate is charged on a column containing Dowex 1X-8 resin in the Cl^- form (100–200 mesh, 4 cm × 20 cm). After the column has been washed with water, the elution is carried out with a 0.1 mole/L potassium chloride solution. A minor orange band descends, followed by a major orange band, each being collected in separate fractions. The eluate of the major band is concentrated to a small volume and filtered. After keeping the filtrate in a refrigerator, crystals of potassium *cis,cis*-diammine-dinitro-oxalatocobaltate(III) hemi-hydrate deposit. They are collected on a filter and recrystallized from warm water (at 40°). The yield is 2.5 g. *Anal.* Calcd. for $K[Co(NO_2)_2(C_2O_4)(NH_3)_2] \cdot 0.5H_2O$: C, 7.48; H, 2.20; N, 17.45. Found: C, 7.69; H, 2.29; N, 17.42.

From the minor fraction, potassium *trans*-(NO_2)-diammine-dinitro-oxalatocobaltate(III) monohydrate is obtainable.[2]

Properties

The *cis,cis*-isomer is rather more stable than the corresponding carbonato complex. The visible absorption spectrum of the *cis,cis*-isomer in water shows ($\tilde{\nu}_{max}$, log ϵ_{max}): (20,900 cm^{-1}, 2.20). The maximum of the $\pi^* \leftarrow \pi$ band is at 29,800 cm^{-1} (log $\epsilon = 3.63$).

Resolution Procedure

To a solution of 3.2 g (0.008 mole) of $(-)_{589}$-bis(1,2-ethanediamine)-dinitrocobalt(III) iodide[5] in 20 mL of water is added 1.3 g (0.008 mole) of silver acetate. The mixture is stirred for 10 minutes at 40°, and the resulting solution is filtered. The precipitated silver halide is washed with 5 mL of warm water (at 50°). The filtrate (combined with the washings) is added with stirring to a solution of 5 g (0.016 mole) of potassium *cis,cis*-diammine-dinitro-oxalatocobaltate hemi-hydrate in 60 mL of warm water (at 35°). Immediately, the orange-yellow diastereoisomeric salt of the $(-)_{589}$oxalato complex begins to deposit. The whole is allowed to stand for half an hour at room temperature. The product is recrystallized twice from warm water in order to increase the optical purity. The recrystallized diastereoisomer is collected on a filter, washed with 50% cold aqueous methanol, ethanol, and diethyl ether, and then dried under vacuum. The yield is 2 g. *Anal.* Calcd. for $[Co(NO_2)_2(en)_2][Co(NO_2)_2(C_2O_4)(NH_3)_2]\cdot H_2O$: C, 12.81; H, 4.30; N, 24.91. Found: C, 13.00; H, 4.48; N, 24.82.

Properties

The CD spectrum in water shows ($\tilde{\nu}_{max}$, $\Delta\epsilon$): (19,500 cm^{-1}, -0.89), (22,300 cm^{-1}, $+1.47$) and (26,500 cm^{-1}, -0.31). (The CD spectrum is measured with a sample converted to the potassium salt by means of the ion-exchange technique described previously.)

Note. The related compounds, *cis*-$K[Co(NO_2)_2(CO_3)(en)]$ and *cis*-$K[Co(NO_2)_2(ox)(en)]$ can be prepared. The former carbonato complex is derived from *cis*-$[Co(CO_3)_2(en)]^-$ and the latter oxalato complex from the carbonato complex.[2]

E. *cis,cis*-DIAMMINECARBONATODI(PYRIDINE)COBALT(III) CHLORIDE MONOHYDRATE

$$cis\text{-}[Co(CO_3)_2(NH_3)_2]^- + 2H^+ + 2py \longrightarrow cis,cis\text{-}[Co(CO_3)(NH_3)_2(py)_2]^+ + CO_2 + H_2O$$

Procedure

To a solution of 9.4 g (0.036 mole) of potassium *cis*-diamminedicarbonatocobaltate(III) in 20 mL of water is added 6 mL (0.072 mole) of pyridine. To the mixture, 60% perchloric acid is added dropwise in an ice bath to adjust the pH to 5.5. After filtration, the filtrate is stirred at room temperature for a while, whereupon pink crystals of *cis,cis*-diamminecarbonatodi(pyridine)cobalt(III) perchlorate deposit. The product is collected on a filter, washed with water, and dissolved in water. The solution is poured into a column of Dowex 50W-X8 resin in the Na^+ form in order to change the counter anion from perchlorate to chloride. The effluent obtained by elution with a 0.3 mole/L sodium chloride is concentrated to 10 mL under reduced pressure at 35°. The concentrated solution is filtered and kept in a refrigerator. Red crystals of *cis,cis*-diamminecarbonatodi(pyridine)cobalt(III) chloride monohydrate deposit. The crude product is recrystallized from warm water at 35°. The crystals are collected on a filter, washed with cold water, ethanol, and diethyl ether, and finally dried under vacuum. The yield is 1.0 g. *Anal.* Calcd. for $[Co(CO_3)(NH_3)_2(py)_2]Cl \cdot H_2O$: C, 36.23; H, 4.98; N, 15.36. Found: C, 36.07; H, 4.93; N, 15.55.

Properties

The visible absorption spectrum in water shows ($\tilde{\nu}_{max}$, log ϵ_{max}); (19,400 cm^{-1}, 2.06) and (27,600 cm^{-1}, 2.07).

Resolution Procedure

Into a solution of 3.3 g (0.009 mole) of *cis*-diamminecarbonatodi(pyridine)cobalt(III) chloride monohydrate in 10 mL of water is poured 1.2 g (0.0038 mole) of sodium $(-)_{546}$-1,2-ethanediamine-bis(oxalato)cobaltate(III) monohydrate[3] with stirring. The solution is cooled in an ice bath, and the sides of the vessel are scratched with a glass rod, whereupon the diastereoisomeric salt of the $(-)_{589}$ carbonato complex deposits. (If the diastereoisomer does not deposit, a small amount of ethanol is added.) The whole is kept in an ice bath for a while. The diastereoisomer is collected on a filter, washed with cold water, and recrystallized several times from water at 35° to increase the optical purity. The recrystallized salt is collected on a filter, washed with cold water, ethanol, and diethyl ether, and dried under vacuum. The yield is 0.3 g. *Anal.* Calcd. for $[Co(CO_3)(NH_3)_2(py)_2][Co(ox)_2(en)] \cdot 2.5H_2O$: C, 31.35; H, 4.49; N, 12.90. Found: C, 31.37; H, 4.67; N, 12.82. The diastereoisomer is separated as before.

Properties

The CD spectrum of $(-)_{589}[Co(CO_3)(NH_3)_2(py)_2]^+$ in water shows ($\tilde{\nu}_{max}$, $\Delta\epsilon$): (18,100 cm^{-1}, -0.59), (20,300 cm^{-1}, 0.92), and (27,800 cm^{-1}, -0.28). (The CD spectrum is measured using a sample with the chloride anion prepared by ion-exchange techniques.) The $(-)_{589}$ isomer in water at 5° does not racemize for at least three days.

Note. A related complex, *cis*-$[Co(CO_3)(py)_2(en)]Cl$ can be prepared from $[Co(CO_3)_2(en)]^-$.[6]

References

1. F. P. Dwyer, I. K. Reid, and F. L. Garvan, *J. Am. Chem. Soc.*, **83,** 1285 (1961).
2. T. Ito and M. Shibata, *Inorg. Chem.*, **16,** 108 (1977).
3. M. Shibata, *Nippon Kagaku Zasshi*, **87,** 771 (1966).
4. Y. Enomoto, T. Ito, and M. Shibata, *Chem. Lett.*, 423 (1974).
5. F. P. Dwyer and F. L. Garvan, *Inorg. Syn.*, **VI,** 195.
6. K. Tsuji, S. Fujinami, and M. Shibata, *Bull. Chem. Soc. Japan*, **54,** 1531 (1981).

III. MONODENTATE GLYCINATO-1,4,7-TRIAZACYCLONONANE COMPLEXES OF COBALT(III)

The chirality of a *fac*(DDD)-[Co(a)(b)(c)(DDD)]-type complex is derived from the arrangement of the unidentates, a, b, and c about the central metal. (DDD refers to a terdentate chelating agent with 3 like donor atoms.) This type of complex has not been prepared previously. Since the ligand, 1,4,7-triazacyclononane, (tacn),[1] is a cyclic terdentate ligand, the chilarity of a [Co(a)(gly)(tacn)]-type complex has the same origin as the *fac*(DDD)-[Co(a)(b)(c)(DDD)]-type complex. In the following, the preparations and resolutions of aquaglycinato(1,4,7-triazacyclononane)cobalt(III), glycinatonitro(1,4,7-triazacyclononane)cobalt(III) and ammineglycinato(1,4,7-triazacyclononane)cobalt(III) complexes are described. The aqua complex is prepared from *mer*(N)-*trans*(NH_3)-diamminecarbonato(glycinato)cobalt(III),[2] and the other complexes are prepared from the aqua complex.

A. AQUAGLYCINATO(1,4,7-TRIAZACYCLONONANE)COBALT(III) PERCHLORATE DIHYDRATE

$$\textit{mer}(\mathrm{N})\text{-}\textit{trans}(\mathrm{NH_3})\text{-}[\mathrm{Co(CO_3)(gly)(NH_3)_2}] + 2\,\mathrm{H^+} + 2\,\mathrm{H_2O} \longrightarrow [\mathrm{Co(gly)(NH_3)_2(H_2O)_2}]^{2+} + \mathrm{CO_2} + \mathrm{H_2O}$$

$$[\mathrm{Co(gly)(NH_3)_2(H_2O)_2}]^{2+} + \mathrm{tacn} \longrightarrow [\mathrm{Co(gly)(H_2O)(tacn)}]^{2+} + \mathrm{H_2O} + 2\mathrm{NH_3}$$

Procedure

Five grams (0.022 mole) of *mer*(N)-*trans*(NH_3)-diamminecarbonato(glycinato)-cobalt(III)[2] is suspended in 20 mL of water, and then 5.0 g (0.022 mole) of tacn·3HCl[1] is added, little by little. The mixture is stirred for a while, whereupon the compound dissolves. To the solution, 60% perchloric acid is added in order to adjust the pH to 2. The corresponding diaqua complex, diamminediaqua-(glycinato)cobalt(III) is formed. The solution is made alkaline with 5 mole/L potassium hydroxide to pH 9, stirred for 15 minutes, and then neutralized with 60% perchloric acid. The resulting solution is filtered and diluted to 2 L with water. This solution is poured into a column containing SP-Sephadex C-25 (4.5 cm × 25 cm, Na^+ form). The adsorbed band is eluted with 0.1 mole/L perchloric acid, and a red band corresponding to the divalent cation is collected. The eluate is concentrated to a small volume under reduced pressure at 35°. A small amount of ethanol and a large amount of diethyl ether are added to the concentrated solution in turn, and then the whole is kept in a refrigerator until red crystals deposit. The crude product is recrystallized from warm water at 60°. The recrystallized crystals are collected on a filter, washed with cold water, ethanol, and diethyl ether, and dried under vacuum. The yield is 2.0 g. *Anal.* Calcd. for $[Co(gly)(H_2O)(tacn)](ClO_4)_2 \cdot 2H_2O$: C, 18.65; H, 4.90; N, 10.88. Found: C, 18.55; H, 4.95; N, 10.62.

Properties

The visible absorption spectrum in water shows ($\tilde{\nu}_{max}$, log ϵ_{max}): (20,000 cm^{-1}, 1.88) and (28,000 cm^{-1}, 1.82).

Resolution Procedure

A solution of 20 mg of this complex in 10 mL of water is poured into a column containing Dowex 50W-X8 resin (200–400 mesh, 2.5 cm × 34.5 cm, Na^+ form). The adsorbed band is eluted with a 0.3 mole/L potassium antimonyl-(+)-tartrate solution. When the adsorbed band descends to four fifths of the column height, the band is washed with water and eluted with 0.3 mole/L sodium perchlorate. The specific rotation of the earlier-effluent is negative at 589 nm.

Properties

The CD spectrum of the $(-)_{589}$-isomer in water shows ($\tilde{\nu}_{max}$, $\Delta\epsilon$): (18,200 cm^{-1}, +0.10), (20,600 cm^{-1}, −0.14), (27,100 cm^{-1}, +0.06) and (30,300 cm^{-1}, −0.01).

B. GLYCINATONITRO(1,4,7-TRIAZACYCLONONANE)COBALT(III) CHLORIDE MONOHYDRATE

$$[Co(gly)(H_2O)(tacn)]^{2+} + NO_2^- \longrightarrow [Co(NO_2)(gly)(tacn)]^+ + H_2O$$

Procedure

To a solution of 1.5 g (0.003 mole) of aquaglycinato(1,4,7-triazacyclononane-cobalt(III) perchlorate dihydrate (see Preparation A) in 20 mL of water, 1 g (0.014 mole) of sodium nitrite is added. The solution is adjusted to pH 4 with 60% perchloric acid and stirred at 50–60° until its color changes from red to yellow. The yellow solution is filtered, diluted to 2 L with water, and poured into a column containing SP-Sephadex C–25 (4.5 cm × 25 cm, Na^+ form). Upon elution with a 0.05 mole/L sodium chloride solution, only one band descends. The eluate is concentrated to a small volume under reduced pressure at 40°, with the simultaneous removal of the deposited sodium chloride. A large amount of ethanol is then added to the concentrate, which is filtered, and kept in a refrigerator to obtain yellow crystals. The crude product is recrystallized from warm water at 50°. The recrystallized crystals are collected on a filter, washed with cold water, ethanol and diethyl ether, and then dried under vacuum. The yield is 0.8 g. *Anal.* Calcd. $[Co(NO_2)(gly)(tacn)]Cl \cdot H_2O$: C, 26.56; H, 5.89; N, 19.37. Found: C, 26.25; H, 5.69; N, 19.47.

Properties

The visible absorption spectrum in water shows ($\tilde{\nu}_{max}$,log ϵ_{max}): (21,800, 2.12).

Resolution Procedure

A solution of 20 mg of this complex in 15 mL of water is poured into a column containing SP-Sephadex C-25 (2.5 cm × 110.0 cm, Na^+ form). When the elution is carried out with a 0.05 mole/L potassium antimonyl-(+)-tartrate solution, the band separates into two bands. When the bands have descended to four fifths of the column height, the column is washed with water. The two bands are eluted with 0.1 mole/L sodium chloride. The specific rotation of the first band is negative at 589 nm.

Properties

The CD spectrum of the $(-)_{589}$ isomer in water shows ($\tilde{\nu}_{max}$, $\Delta\epsilon$): (19,800 cm^{-1}, −0.32), (21,300 cm^{-1}, +0.15) and (23,600 cm^{-1}, −0.66).

C. AMMINEGLYCINATO(1,4,7-TRIAZACYCLONONANE)COBALT(III) IODIDE MONOHYDRATE

$$[Co(gly)(H_2O)(tacn)]^{2+} + NH_3 \longrightarrow [Co(gly)(NH_3)(tacn)]^{2+} + H_2O$$

Procedure

One gram (0.003 mole) of aquaglycinato(1,4,7-triazacyclononane)cobalt(III) perchlorate dihydrate is dissolved in 100 mL of 28% aqueous ammonia. The solution is allowed to stand in a closed flask at 60° until it turns orange. The resulting solution is charged on a column containing Dowex 50W-X8 resin (4.5 cm × 10 cm, 100–200 mesh, Na^+ form). By elution with 0.3 mole/L sodium iodide, one orange band descends. The band is collected, concentrated to a small volume under reduced pressure at 40°, and filtered. The filtrate is kept in a refrigerator until orange crystals deposit. They are recrystallized from warm water at 60°. The recrystallized product is collected on a filter, washed with cold water, ethanol, and diethyl ether, and dried under vacuum. The yield is 1.0 g. *Anal.* Calcd. for $[Co(gly)(NH_3)(tacn)]I_2 \cdot H_2O$: C, 17.31; H, 4.37; N, 12.62. Found: C, 17.37; H, 4.26; N, 12.52.

Properties

The visible absorption spectrum in water shows ($\tilde{\nu}_{max}$, log ϵ_{max}): (20,800 cm^{-1}, 1.96) and (29,100 cm^{-1}, 2.00).

Resolution Procedure

A solution of 20 mg of this complex in 20 mL of water is charged on a column containing SP-Sephadex C-25 (2.5 cm × 110 cm, Na^+ form). When the elution is carried out with 0.05 mole/L sodium antimonyl-(+)-tartrate, two bands descend. When the earlier band has descended to four fifths of the column height, the bands are washed with water and eluted with 0.1 mole/L sodium chloride. The specific rotation of the earlier band is negative at 589 nm.

Properties

The CD spectrum of the $(-)_{589}$ isomer in water shows ($\tilde{\nu}_{max}$, $\Delta\epsilon$): (18,600 cm^{-1}, −0.02), (19,800 cm^{-1}, +0.06) and (22,000 cm^{-1}, −0.42).

Note. The related complexes, $[Co(CN)(gly)(tacn)]^+$, $[Co(NCS)(gly)(tacn)]^+$, $[Co(OH)(gly)(tacn)]^+$, $[Co(N_3)(gly)(tacn)]^+$, and $[Co(Cl)(gly)(tacn)]^+$ have been prepared and resolved.[3] The absolute configuration of the $(-)_{589}[Co(gly)(NH_3)(tacn)]I_2 \cdot H_2O$ was determined by X-ray analysis[4] and those of the other complexes by derivatives from the optically active complexes, $(-)_{589}[Co(gly)(H_2O)(tacn)]^{2+}$ and $(-)_{589}[Co(NCS)(gly)(tacn)]^+$.[3]

References

1. J. E. Richman and T. J. Atkins, *J. Am. Chem. Soc.*, **96,** 2268 (1974).
2. S. Kanazawa and M. Shibata, *Bull. Chem. Soc. Japan,* **44,** 2424 (1971).
3. S. Shimba, S. Fujinami, and M. Shibata, *Bull. Chem. Soc. Japan,* **53,** 2523 (1980).
4. S. Sato, S. Ohba, S. Shimba, S. Fujinami, M. Shibata, and Y. Saito, *Acta Cryst.*, **B36,** 43 (1980).

16. THREE ISOMERS OF THE *trans*-DIAMMINE-[N,N′-BIS(2-AMINOETHYL)-1,2-ETHANEDIAMINE]-COBALT(III) COMPLEX CATION

Submitted by S. UTSUNO,* Y. SAKAI,* Y. YOSHIKAWA,† and H. YAMATERA†
Checked by B. WANG and Y. SASAKI‡

The title complex (as the chloride) was first reported by Kyuno and Bailar,[1] who obtained *cis*-α and *cis*-β isomers by the ligand displacement reaction of *cis*-α-dichloro(N,N′-bis(2-aminoethyl)-1,2-ethanediamine)cobalt(III) chloride in gaseous and liquid ammonia.§ However, its *trans* isomers were not prepared until 1980.[2]

*Department of Chemistry, Faculty of Science, Shizuoka University, Oya, Shizuoka 422, Japan.
†Department of Chemistry, Faculty of Science, Nagoya University, Chikusa-ku, Nagoya 464, Japan.
‡Department of Chemistry, Faculty of Science, Tohoku University, Sendai 980, Japan.
§N,N′-(2-aminoethyl)-1,2-ethanediamine = triethylenetetramine = trien.

The $[CoX_2(trien)]$-type complexes have a marked tendency to prefer *cis* to a *trans* configuration, probably because of the ring strain in the latter isomer. Among the known *trans*-$[CoX_2(trien)]$-type complexes (X = Cl,[3] CN,[4] CH_3CO_2,[5] $C_2H_5CO_2$,[5] $ClCH_2CO_2$[5]), the dicyano complex is the only one that was obtained in the *meso* form, and $(+)_{589}$-*trans*-$[CoCl_2(trien)]$ is the only one that was isolated as an optical isomer. The present diammine complex is the first example of this type of complex, for which an entire set of the three *trans* isomers of the (+)-, (−)-, and *meso*-types has been isolated.

The *trans*-$[Co(NH_3)_2(trien)]^{3+}$ complex cannot be prepared by the ligand displacement of *cis*- or *trans*-$[CoCl_2(trien)]^+$ with ammonia. The method presented here includes the conversion[6] of coordinated isothiocyanate ions of *trans*-$[Co(NCS)_2(trien)]^+$,[7] and the subsequent separation of the three isomers using columns of SP-Sephadex C-25.

$$[Co(NCS)_2(trien)]^+ + 8H_2O_2 \longrightarrow [Co(NH_3)_2(trien)]^{3+} + 2H^+ + 2SO_4^{2-} + 2CO_2 + 4H_2O$$

Procedure

■ **Caution.** *Danger! Perchlorate salts of these isomers may explode. The solid perchlorates should be dried at room temperature. The washings which contain perchlorates and organic solvents should not be heated. All reactions should be carried out in an appropriate hood.*

A solution of 210 mg (0.5 mmole) of *trans*-bis(thiocyanato-*N*)(trien)cobalt(III) perchlorate[7] in 40 mL of 0.1 *M* HCl is warmed to 40°, and 2.5 mL of 30% H_2O_2 is added to it. The mixture is allowed to stand for four days at 40°.* The resulting orange-yellow solution is diluted with 100 mL of water and is passed through a column (20 mm × 70 mm long) of SP-Sephadex C-25 (H^+ form). The column is washed with 500 mL of 0.03 *M* HCl, and then with 250 mL of 0.1 *M* HCl. At first, a trace amount of a reddish-orange species moves down, and then the adsorbed species separate into an orange-yellow band (faster) and a yellow band (slower).** They are eluted with 0.5 *M* HCl, and the yellow eluate is concentrated by a rotatory vacuum evaporator almost to dryness.† This liquid and its washings are transferred to a 30-mL beaker. To the combined solution, absolute ethanol, amounting to about 10 times the volume of the solution, is

*When the pure racemic isomer is used as a starting material, the reaction is practically complete in about 40 hours at 40°.

**The orange-yellow band may be composed of $[Co(NCS)(NH_3)(trien)]^{2+}$, but this has not been confirmed. The checkers report that a small amount of a pink band forms first.

†Sometimes a red species is formed, which is soluble in 90% ethanol. The checkers report that formation of this species can be avoided if the solution is kept acidic.

added in small portions.* This causes the separation of a yellow solid, which is collected on a sintered-glass filter and washed with ethanol and diethyl ether. The checkers report a yield of 55 mg from 210 mg of the starting material.

A. SEPARATION OF THE *MESO* AND *RACEMIC* ISOMERS

One hundred milligrams of the above product, dissolved in 0.5 mL of 0.03 *M* HCl, is placed on the top of a column (30 mm × 140 cm long) of SP-Sephadex C-25, and eluted with 0.15 *M* Na_2SO_4 containing 5 mL of 1 *M* H_2SO_4 per 1000 mL of the solution. The two yellow bands are collected separately. The first band consists of the *meso* isomer, and the second, of the *racemic* isomer. Each solution is diluted with about 8 L of 0.01 *M* HCl (about 20 times the eluate), and is allowed to pass through a column (30 mm × 70 mm long) of SP-Sephadex C-25. After the column is washed with 1000 mL of 0.03 *M* HCl, the complex is eluted with 1 *M* HCl, and the eluate is treated as described in the previous paragraph. The checkers report that λ_{max} is at 460 nm for the *meso* isomer, and at 468 for the *racemic* form. *Anal.* Calcd. for $[Co(NH_3)_2(C_6H_{18}N_4)]Cl_3 \cdot H_2O$: C, 19.82; H, 7.21; N, 23.11; Co, 16.21. Found for the *meso* isomer: C, 19.77; H, 6.96; N, 23.08; Co, 16.03. The perchlorate of each isomer is easily obtained by the addition of 70% $HClO_4$ to a concentrated aqueous solution of the chloride.

B. OPTICAL RESOLUTION OF THE *RACEMIC* ISOMER

A solution of 100 mg of the *racemic-trans*-diammine(trien)cobalt(III) perchlorate in 0.5 mL of 0.03 *M* HCl is treated in a manner similar to that described above for the separation of the *meso* and *racemic* isomers, except that the eluent used is 0.15 *M* sodium (+)-tartratoantimonate(III) instead of Na_2SO_4. The first band consists of $(-)_{546}$-$[Co(NH_3)_2(trien)]^{3+}$, and the second, the (+)-isomer. *Anal.* Calcd. for $[Co(NH_3)_2(C_6H_{18}N_4)](ClO_4)_3 \cdot H_2O$: C, 12.97; H, 4.72; N, 15.13; Co, 10.61. Found for the (+)-isomer: C, 13.00; H, 4.47; N, 15.01; Co, 10.60. Found for the (−)-isomer: C, 12.89; H, 4.49; N, 15.01; Co, 10.46. $[\alpha]_{546}$ = +360; and −358 (0.2% in 0.01 *M* $HClO_4$).

Properties

All three isomers of *trans*-diammine(trien)cobalt(III) chloride (or perchlorate) are very soluble in water but sparingly soluble in ethanol. Although the solid complex salts are stable for over a year when stored in a refrigerator (5°), they

*The complex chloride separates as a fine powder if ethanol is added all at once.

change from yellow to red when stored for months at room temperature (25–30°). They are stable in acidic solutions, but unstable in neutral solutions. In order to prevent the displacement of the coordinated ammonia and the isomerization to other *trans* isomers, the solution of each isomer must be acidified.

References

1. E. Kyuno and J. C. Bailar, Jr., *J. Am. Chem. Soc.*, **88,** 1124 (1966).
2. S. Utsuno, Y. Sakai, Y. Yoshikawa, and H. Yamatera, *J. Am. Chem. Soc.*, **102,** 6903 (1980).
3. (a) A. M. Sargeson and G. H. Searle, *Inorg. Chem.*, **6,** 787 (1967); (b) D. A. Buckingham, P. A. Marzilli, and A. M. Sargeson, *Ibid.*, **6,** 1032 (1967).
4. (a) E. A. Spatola, *Diss. Abst. Int. B*, **32,** 808 (1971); (b) R. K. Wismer and R. A. Jacobson, *Inorg. Chim. Acta*, **7,** 477 (1973).
5. K. Kuroda and P. S. Gentle, *J. Inorg. Nucl. Chem.*, **29,** 1963 (1967).
6. S. M. Caldwell and A. R. Norris, *Inorg. Chem.*, **7,** 1667 (1968).
7. Y. Sakai, Y. Nokami, H. Kanno, and S. Utsuno, *Chem. Lett.*, **1981,** 371.

17. 1,4,8,11-TETRAAZACYCLOTETRADECANE-5,7-DIONE (*dioxocyclam*) COPPER(II) AND AQUEOUS COPPER(III) COMPLEXES

Submitted by LUIGI FABBRIZZI,* ANTONIO POGGI,* and BARBARA SEGHI*
Checked by D. B. MORSE and D. H. BUSCH†

This 14-membered macrocycle is an intermediate[1] in the synthesis of the more celebrated tetramine analog *cyclam*, which is more easily prepared by a template route.[2] Recently, it has been discovered that *dioxocyclam*, in the presence of aqueous Cu(II), loses two protons (from the amido groups) and simultaneously incorporates the metal ion, to give a neutral dioxocyclamato complex, which is stable over a large pH range.[3] Moreover, it has been found that the Cu(II) dioxocyclamato complex can be oxidized in aqueous solution (chemically or anodically), to produce an authentic Cu(III) complex, which persists in water for hours.[4] It should be noticed that the corresponding copper(III) complex with the tetramine analog, $[Cu(cyclam)]^{3+}$, cannot be generated anodically in water, whereas, in anhydrous acetonitrile, it has a lifetime of only a few seconds.[5] The dioxocyclamato ligand represents the prototype of a unique class of macrocycles

*Department of General Chemistry, University of Pavia, Pavia 27100, Italy.
†Department of Chemistry, The Ohio State University, Columbus, OH 43210.

that allow preparation of aqueous solutions of complexes of the otherwise very unstable copper(III).

A. 1,4,8,11-TETRAAZACYCLOTETRADECANE-5,7-DIONE (*dioxocyclam*)

$$CH_2(COOC_2H_5)_2 + H_2N(CH_2)_2NH(CH_2)_3NH(CH_2)_2NH_2 \longrightarrow \text{dioxocyclam} + 2C_2H_5OH$$

Procedure

Diethyl malonate, 15 g (0.093 mole), purified by distillation under reduced pressure, and 15 g (0.093 mole) of $H_2N(CH_2)_2NH(CH_2)_3NH(CH_2)_2NH_2$, conveniently prepared according to the procedure described in reference 2, are dissolved in 1900 mL of absolute ethanol, contained in a 3-L round-bottomed flask, to give a 5×10^{-2} *M* solution, which is refluxed for five days. The yellow solution is evaporated to about 100 mL on a rotary evaporator and is allowed to stand at room temperature in a corked bottle. In two to three days or less, white crystals form, which are removed by filtration, washed with small portions of cold ethanol and diethyl ether, and dried under vacuum. The yield is 5.5 g (24%). *Anal.* Calcd. for $C_{10}H_{20}N_4O_2$: C, 52.61; H, 8.83; N, 24.55. Found: C, 52.30; H, 8.87; N, 24.77.

Properties

Dioxocyclam is a white, crystalline material that can be recrystallized from ethanol. Melting point: 176–177°. It dissolves in water, in which it behaves as a diprotic base (log K_1 ($B + H^+ = BH^+$) = 9.57; log K_2 ($BH^+ + H^+ = BH_2^{++}$) = 5.97).[3] The protonation processes involve the two amine nitrogen atoms. *Dioxocyclam* is a weaker base than *cyclam*, especially as far as the second step is concerned (log K_1 = 11.50; log K_2 = 10.30), due to the shorter distance between the amine centers (which in *cyclam* are in *trans* positions) and to the consequently stronger electrostatic repulsions.

B. 1,4,8,11-TETRAAZACYCLOTETRADECANE-5,7-DIONATO(2−)-COPPER(II), [Cu(dioxocyclamato(2−))]

$$C_{10}H_{20}N_4O_2 + Cu(OOCCH_3)_2 \cdot H_2O \longrightarrow$$

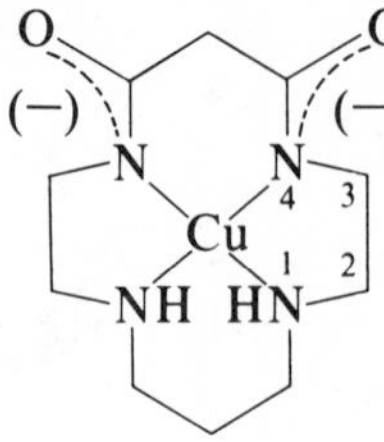

$$+ 2\ CH_3COOH + H_2O$$

Procedure

Dioxocyclam (1.00 g, 4.4 mmoles) is dissolved in 20 mL of 95% ethanol. To this solution 0.88 g of copper(II) acetate monohydrate, dissolved in 60 mL of 95% ethanol,* are added with stirring. A pink-violet, pearly precipitate forms immediately, which is removed by filtration, washed with diethyl ether, and dried under vacuum overnight. The yield is 1.30 g (91%). *Anal.* Calcd. for $C_{10}H_{18}N_4O_2Cu \cdot 2H_2O$: C, 36.86; H, 6.81; N, 17.20. Found: C, 37.50; H, 6.51; N, 17.48.

Properties of the Copper(II) Complex and Formation of the Aqueous Copper(III) Complex Species

The copper(II) dioxocyclamato complex is moderately soluble in water, slightly soluble in methanol and ethanol, and insoluble in most organic solvents. The visible spectrum of the aqueous solution presents an absorption band centered at 19,850 cm^{-1} (molar absorptivity: 101 L $mole^{-1}cm^{-1}$), typical of a tetragonally distorted octahedral complex, in which the macrocycle chelates in a coplanar fashion. The corresponding $[Cu(cyclam)]^{2+}$ complex in water presents a band at 19,900 cm^{-1}. It has been demonstrated that, for tetragonal tetramine complexes, the energy of the absorption band is linearly related to the Cu^{II}-N in-plane bond strength.[6] Consequently, the cyclam and dioxocyclamato(2−) ligands appear to exert coordinative interactions of approximately the same intensity.

The aqueous copper(III) dioxocyclamato complex can be prepared by addition

*The checkers report that the copper(II) acetate must be placed into a 50:50 ethanol:water solution, and the filtrate of this solution is added to 20 mL of a 95% ethanol solution of the ligand; the undissolved copper(II) acetate must be washed with a 50:50 ethanol:water solution several times.

of solid $K_2S_2O_8$ to an aqueous solution of the copper(II) complex; the pink-violet Cu(II) solution slowly turns yellow-brown due to the formation of the Cu(III) complex. The authenticity of the trivalent state can be demonstrated through EPR studies. On oxidation, the signal due to the Cu(II) complex disappears, as is expected for the formation of a d^8, esr-silent chromophore. The aqueous $[Cu(dioxocyclamato(2-))]^+$ complex is not stable indefinitely, but decomposes after few hours at room temperature. Upon addition of conventional reducing agents (e.g., $Na_2S_2O_3$ or Na_2SO_3) to the yellow-brown solution of the trivalent complex, the pink-violet color of the divalent species is immediately restored, demonstrating reversible redox behavior in this system. The aqueous $[Cu^{III}(dioxocyclamato(2-))]^+$ complex can also be generated through controlled potential electrolysis, using a conventional three-electrode cell and imposing on the platinum gauze working electrode a potential of about +0.8 v with respect to the reference calomel electrode. The standard electrode potential for the electrochemically reversible $[Cu^{III}(dioxocyclamato(2-))]^+/[Cu^{II}(dioxocyclamato(2-))]$ redox couple is +0.88 v vs SHE.[4,7]

References

1. I. Tabushi, Y. Taniguchi, and H. Kato, *Tetrahedron Letters*, 1049 (1977).
2. E. K. Barefield, F. Wagner, A. W. Herlinger, and A. R. Dahl, *Inorg. Synth.*, **16**, 220 (1975).
3. M. Kodama and E. Kimura, *J. Chem. Soc., Dalton Trans.*, 325 (1979).
4. L. Fabbrizzi and A. Poggi, *J. Chem. Soc., Chem. Comm.*, 646 (1980).
5. L. Fabbrizzi et al., *J. Chem. Soc., Dalton Trans.*, in press.
6. L. Fabbrizzi, P. Paoletti, and A. B. P. Lever, *Inorg. Chem.*, **15**, 1502 (1976).
7. M. Kodama and E. Kimura, *J. Chem. Soc., Dalton Trans.*, 694 (1981).
8. L. Fabbrizzi, A. Poggi, and P. Zanello, *J. Chem. Soc., Dalton Trans.*, 2191 (1983).

18. CHIRAL ARENE-TRICARBONYLCHROMIUM COMPLEXES: RESOLUTION OF ALDEHYDES

Submitted by A. SOLLADIE-CAVALLO, G. SOLLADIE and E. TSAMO*
Checked by J. W. FALLER and K. H. CHAO†

Functionalized chiral arene-tricarbonylchromium complexes are of great potential interest in asymmetric synthesis. Of these chiral complexes only the racemic acids *2* have been readily resolved previously.[1–4] Resolution of the racemates of

*Laboratoire de Chimie Organique de l'Ecole Nationale Superieure de Chimie, Universite Louis Pasteur, BP 296/R8, Strasbourg, France.

†Department of Chemistry, Yale University, New Haven, CT 06511.

aldehydes *1*a, b, c, and d has been achieved[5] in three straightforward steps by chromotography of the derivatives *4*, prepared by reaction of the aldehydes with (S)-(−)-5-(α-methylbenzyl)semioxamazide *3*.[6]

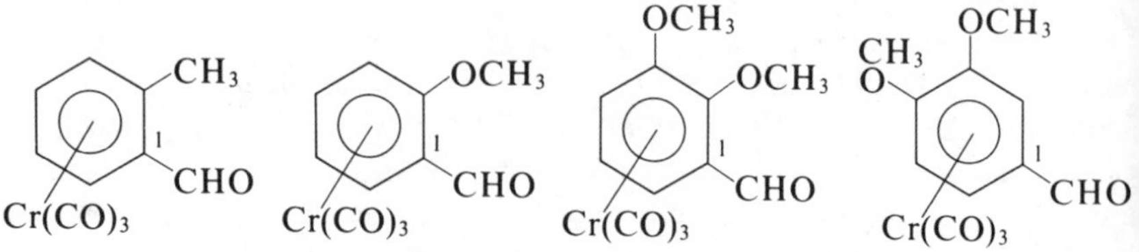

CH_3 CHO 1 $Cr(CO)_3$ ***1a***

OCH_3 CHO 1 $Cr(CO)_3$ ***1b***

OCH_3 OCH_3 CHO 1 $Cr(CO)_3$ ***1c***

OCH_3 CH_3 O CHO 1 $Cr(CO)_3$ ***1d***

R COOH 1 $Cr(CO)_3$ ***2***

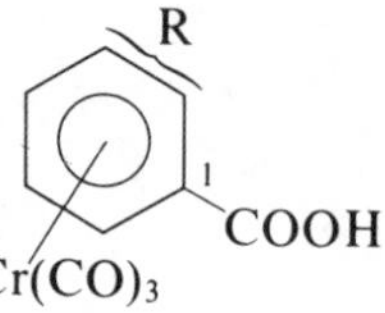

1. Preparation of the semioxamazone

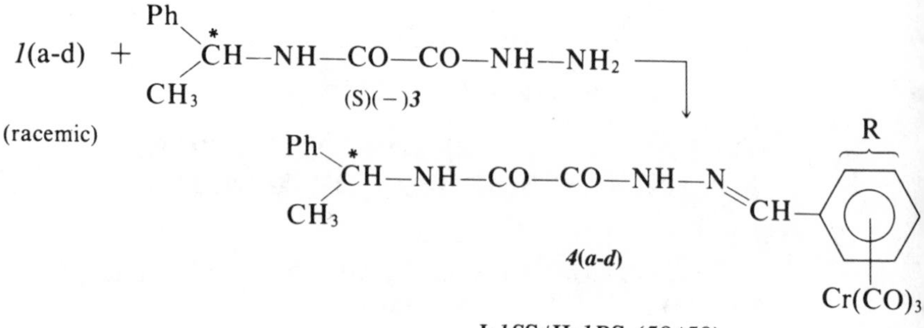

1(a-d) (racemic) + Ph(CH$_3$)C*H—NH—CO—CO—NH—NH$_2$ (S)(−)*3* →

Ph(CH$_3$)C*H—NH—CO—CO—NH—N=CH—C$_6$H$_4$(R)Cr(CO)$_3$

4(a-d)

I:*1*SS/II:*1*RS (50/50)

2. Separation of diastereomeric mixtures by chromatography, e.g., *4a*I from *4a*II; and

3. Hydrolysis of the diastereomers, for example:

$$4aI \xrightarrow[60\%\ H_2SO_4]{\text{toluene/benzene}} \text{a pure enantiomer of } 1a$$

Procedures

■ **Caution.** *Benzene, a reported carcinogen, is used in several steps in these syntheses. Appropriate precautions must be taken, and an efficient hood must be used.*

A. SYNTHESIS OF THE SEMIOXAMAZONES (*4a–d*)

The (S)-(−)-5-(α-methylbenzyl)semioxamazide *3* (mp 168–169°, $[\alpha]_D$ −103°, c = 1, $CHCl_3$) is prepared by sequential addition of (S)-(−)-α-phenethylamine* and hydrazine to diethyl oxalate.[6]

The substituted benzentricarbonylchromium derivatives were prepared by protection of the aldehyde, heating in the presence of $Cr(CO)_6$, and subsequent deprotection of the aldehyde.[5]†

The semioxamazone is prepared by refluxing a mixture of 0.04 mole of *3*, 0.04 mole of aldehyde *1*, and 0.5 g of *p*-toluenesulfonic acid in 400 mL of anhydrous benzene until no starting aldehyde can be detected by TLC (about 10 minutes for *4b*). Pre-coated TLC plates (silica gel 60 F-254, eluent: diethyl ether/hexane 90/10). The benzene is removed on a rotary evaporator and a red-orange, solid crude product is obtained, which is a ~50:50 mixture of the diastereomers.

*The checkers used Aldrich gold label (R)-(+)-α-phenethylamine and obtained $[\alpha]_D$ +110° for the (R) semioxamazide.

†The checkers found the published description of the preparation rather abbreviated, and expanded the procedure to provide more detail, for the example of (*o*-anisaldehyde)tricarbonylchromium.

Protection of the Aldehyde. A mixture of *o*-anisaldehyde (13.6 g, 100 mmoles), ethylene glycol (7.45 g, 120 mmoles), and *p*-toluenesulfonic acid (0.5 g) is refluxed with benzene in a Dean-Stark apparatus until no starting aldehyde can be detected by TLC (~6 hr). After being cooled, the benzene solution is washed three times with 5% aqueous $NaHCO_3$, then with saturated salt water until the washings are neutral, and then dried over Na_2SO_4. The solvent is removed under vacuum and the product collected in 96% yield by vacuum distillation (80°, 1 torr Hg).

Preparation of the Complex. Following a published procedure,[7] 4 g (18 mmoles) $Cr(CO)_6$, 19 mL (95 mmol) of the protected aldehyde, 120 mL freshly distilled dibutyl ether, and 10 mL tetrahydrofuran are placed in a 250 mL round-bottomed flask fitted with a nitrogen inlet and a simple reflux condenser. The mixture is heated to reflux for 24 hours. The yellow solution is cooled and filtered, and the solvents are removed by vacuum distillation. The residue is chromatographed on a silica gel column (4 cm × 20 cm), and a yellow band collected upon elution with a mixture of ether and pentane (70:30). The solvents are removed on a rotary evaporator and yellow crystals are collected (4.24 g, 80%).

Deprotection. The complex (4.24 g, 14 mmoles) is dissolved in 20 mL of absolute ethanol and 2 mL of 1 *N* HCl is added to the solution. The mixture is stirred until no starting complex can be detected by TLC (10 min), and the color of the solution changes from yellow to red-orange. The solution is extracted with dichloromethane. The methylene chloride layer is separated and washed three times with 5% aqueous Na_2CO_3 and then with saturated NaCl solution, until the washings are neutral. After drying over Na_2SO_4, the solvent is removed on a rotary evaporator and the residue crystallized from a mixture of dichloromethane and pentane to yield 3.24 g of red crystals (90%).

B. SEPARATION OF DIASTEREOMERS

The separations are performed by column chromatography using low pressure techniques.[8] The use of a 400 mm × 30 mm silica gel column along with elution with 80% ether and 20% pentane allows the separation of 1 g of the diastereomeric mixture. The purity of the fractions can be followed conveniently by TLC.*

C. HYDROLYSIS OF THE PURE DIASTEREOMERIC SEMIOXAMAZONES

The pure semioxamazone (0.07 mole) is refluxed in 100 mL of a 60/40 mixture of toluene/benzene and 50 mL of 60% (by volume) H_2SO_4.† The progress of the reaction can be followed conveniently by TLC. After cooling the solution, the aqueous phase is decanted and the organic phase is washed three times with 50 mL of 10% sodium carbonate and then with 20 mL of water until the washings are neutral. The organic layer is dried over sodium sulfate, the solvent evaporated, and the crude aldehyde product chromatographed on silica gel.

Properties

1. Semioxamazone *4a*

Yield: 80%. Chromatography solvent: diethyl ether/petroleum ether (EE/PE) 90/10.

Diastereomer I (*4a* I). mp 155–156°; R_F 0.75 (EE/PE 90/10); IR($CHCl_3$): ν_{NH} 3360, 3290, $\nu_{C=O}$ 1975, 1900, ν_{amide} 1680 cm^{-1}; ^1H-nmr($CDCl_3$/TMS, 250 MHz) δ: 1.54(d, 3H, J = 7Hz), 2.26(s, 3H), 4.98(d, 1H), 5.12(t, 1H), 5.43(t, 1H), 6.17(d, 1H arom.), 5.00(m, 1H), 7.17(5H arom.), 7.57(d, 1H, NH), 8.02(s, 1H imine), 10.12(s, 1H, NH); *Anal.* Calcd. for $C_{21}H_{19}CrN_3O_5$: C, 56.63; H, 4.30. Found: C, 56.80; H, 4.39. $[\alpha]_D$ +1013° (c 0.76, $CHCl_3$).

Diastereomer II (*4a* II). mp 166–168°; R_F 0.46, (EE/PE 90/10); IR($CHCl_3$) and nmr($CDCl_3$/TMS) are identical to those of diastereomer *4a* I; $[\alpha]_D$ −1195° (c 0.69, $CHCl_3$).

*The checkers found that a 300-mm gravity column effected adequate separation using Mallinkrodt Silicar CC-7.

†The checkers found improved yields when the hydrolyses are carried out at room temperature. For example, a pure diastereomer, *4b*I, is hydrolyzed by dissolving in benzene (1 g/20 mL) and stirring with 10 mL of 60% H_2SO_4. After 20 minutes no more starting material can be detected by TLC.

2. Semioxamazone *4b*

Yield: 90%; chromatography solvent: diethyl ether.

Diastereomer I (*4b* I). mp >190° (decomposes); R_F 0.75 (ether); IR($CHCl_3$): ν_{NH} 3380, 3300, $\nu_{C=O}$ 1975, 1905, ν_{amide} 1680 cm^{-1}; ^{1}H-nmr($CDCl_3$/TMS 250 MHz) δ: 1.60(d, 3H, J ≃ 7), 3.81(s, 3H), 4.98(t, 1H arom.), 5.09(d, 1H arom.), 5.12(m, 1H), 5.70(t, 1H arom.), 6.54(d, 1H arom.), 7.35(s, 5H arom.), 7.78(d broad, 1H NH), 8.28(s, 1H), 10.4(s, 1H NH); *Anal.* Calcd. for $C_{21}H_{19}CrN_3O_6$: C, 54.66 H, 4.12 N, 9.11. Found: C, 54.86; H, 4.35; N, 9.30. $[\alpha]_D$ +986° (c 0.07, $CHCl_3$).*

Diastereomer II (*4b* II). mp 99–101°; R_F 0.57 (diethyl ether); IR($CHCl_3$) and nmr ($CDCl_3$/TMS) are identical to those of diastereomer *4b*I; $[\alpha]_D$ −959° (c 0.07, $CHCl_3$).*

3. Semioxamazone *4c*

Yield: 70%; chromatography solvent: diethyl ether.

Diastereomer I (*4c* I). mp 96–98°; R_F 0.72 (diethyl ether); IR($CHCl_3$): ν_{NH} 3380, 3310, $\nu_{C=O}$ 1970, 1900, ν_{amide} 1680 cm^{-1}; ^{1}H nmr($CDCl_3$/TMS 250 MHz) σ: 1.50(d, 3H), 3.75(s, 3H), 3.85(s, 3H), 5.20(m, 1H + 2H arom.), 5.75(d, 1H arom.), 7.25(s, 5H arom.), 7.65(d, 1H NH), 8.30(s, 1H), 10.45(s broad, 1H NH); *Anal.* Calcd. for $C_{22}H_{21}CrN_3O_7$: C, 53.77; H, 4.28; N, 8.55. Found: C, 53.67; H, 4.31; N, 8.55. $[\alpha]_D$ +556° (c 0.06, $CHCl_3$).

Diastereomer II (*4c* II). mp 118–120°; R_F 0.57 (diethyl ether); IR($CHCl_3$) and ^{1}H nmr ($CDCl_3$/TMS) are identical to those of diastereomer *4c* I; $[\alpha]_D$ −415° (c 0.05 $CHCl_3$).

4. Semioxamazone *4d*

Yield: 70%; chromatography solvent: diethyl ether.

Diastereomer I (*4d* I). mp 108–110°; R_F 0.59 (diethyl ether); IR($CHCl_3$): ν_{NH} 3380, 3300, $\nu_{C=O}$ 1970, 1895, ν_{amide} 1680 cm^{-1}; ^{1}H nmr($CDCl_3$/TMS 250 MHz) δ: 1.60(d, 3H), 3.75(s, 6H), 5.00(m, 1H), 5.15(s, 2H arom.), 5.90(s, 1H arom.), 7.20(s, 5H arom.), 7.75(broad, 1H NH), 7.80(s, 1H), 10.90(s broad, 1H NH). *Anal.* Calcd. for $C_{22}H_{21}CrN_3O_7$: C, 53 .77; H, 4.28; N, 8.55. Found: C, 53.85; H. 4.38; N, 8.59. $[\alpha]_D$ +450° (c 0.09, $CHCl_3$).

*The checkers resolved the isomers of the opposite configuration and found the mp for *4b*I to be 184–186° (dec) and $[\alpha]_D$ −1120°. The mp for *4b*II was 99–101° and $[\alpha]_D$ +1080°. At 500 MHz three resonances were sufficiently resolved to distinguish I and II respectively: δ 8.287, 8.297; 3.835, 3.833; and 1.604, 1.602. The checkers also observed that decomposition occurs slowly in chloroform, which leads to a slow decrease in rotation with time.

Diastereomer II (*4d* II). mp 105–107°; R_F 0.39 (diethyl ether); IR($CHCl_3$) and ^1H nmr ($CDCl_3$/TMS) are identical to those of diastereomer *4d* I; $[\alpha]_D$ −479 (c 0.10 $CHCl_3$).

5. Optically Active Complexes

1a: mp 99–100°; R_F 0.95 (diethyl ether/hexane 9/1); IR($CHCl_3$): ν_{CHO} 2860, 2720 (VW), $\nu_{C=O}$ 1970, 1895, ν_{CO} 1680 cm^{-1}; ^1H nmr($CDCl_3$/TMS) δ: 2.45(s, 3H), 4.95(d, 1H arom.), 5.12(t, 1H arom.), 5.62(t, 1H arom.), 5.95(d, 1H arom.), 9.72(s, 1H).

From *4a* I: $[\alpha]_D$ +665°, (c 0.22, $CHCl_3$). From *4a* II: $[\alpha]_D$ −664°, (c 0.26, $CHCl_3$); (lit.[3] $[\alpha]_D$ −660°).

1b: mp 98–99°; R_F 0.89 (diethyl ether); IR($CHCl_3$): ν_{OMe}, ν_{CHO} 2840, 2810 (W), $\nu_{C=O}$ 1975, 1900, ν_{CO} 1675 cm^{-1}; ^1H nmr($CDCl_3$/TMS) δ: 3.80(s, 3H), 4.95(t, 1H arom.), 5.0(d, 1H arom.), 5.75(t, 1H arom.), 6.15(d, 1H arom.), 10.1(s, 1H).

From *4b* I: $[\alpha]_D$ +1015°, (c 0.06, $CHCl_3$). From *4b* II: $[\alpha]_D$ −1020°, (c 0.09, $CHCl_3$); (Lit.[9] $[\alpha]_D$ −1000°).*

1c: mp 59–61°; R_F 0.94 (diethyl ether); IR($CHCl_3$): ν_{OMe}, ν_{CHO} 2860, 2820 (W), $\nu_{C=O}$ 1975, 1900, ν_{CO} 1680 cm^{-1}; ^1H nmr($CDCl_3$/TMS) δ: 3.75(s, 3H), 3.95(s, 3H), 5.10(t, 1H arom.) 5.45(d, 1H arom.), 5.60(d, 1H arom.), 10.05(s, 1H).

From *4c* I: $[\alpha]_D$ +360°, (c 0.12, $CHCl_3$). From *4c* II: $[\alpha]_D$ −387°, (c 0.19, $CHCl_3$).

1d: mp 83–85°; R_F 0.93 (diethyl ether); IR($CHCl_3$): ν_{OMe} 2810, ν_{CHO} 2810, 2720 (vw), $\nu_{C=O}$ 1970, 1885 ν_{CO} 1675 cm^{-1}; ^1H nmr($CDCl_3$/TMS) δ: 3.75(s, 3H), 3.80(s, 3H), 5.15(d,1H arom.), 5.55(d, 1H arom.), 5.85(s, 1H arom.), 9.30(s, 1H).

From *4d* I: $[\alpha]_D$ +793°, (c 0.60, $CHCl_3$). From *4d* II: $[\alpha]_D$ −818°, (c 0.77, $CHCl_3$).

References

1. A. Mandelbaum, Z. Neuwirth, and M. Caïs, *Inorg. Chem.*, **2,** 902 (1963).
2. R. Dabard and A. Meyer, *C. R. Acad. Sci. Paris,* **264C,** 903 (1967).
3. R. Dabard, A. Meyer, and G. Jaouen, *C. R. Acad. Sci. Paris,* **268C,** 201 (1969).
4. H. Falk, K. Schlögl, and W. Steyrer, *Monatsh. Chem.*, **97,** 1029 (1966).
5. A. Solladié-Cavallo, G. Solladié, and E. Tsamo, *J. Org. Chem.*, **44,** 4189 (1979).
6. N. J. Leonard and J. H. Boyer, *J. Org. Chem.*, **15,** 42 (1950).
7. C. A. L. McHaffy and P. L. Pauson, *Inorg. Syn.*, **19,** 154 (1979).
8. W. C. Still, M. Kahn, and A. Mitra, *J. Org. Chem.*, **43,** 2923 (1978).
9. R. Dabard and G. Jaouen, *Tetrahedron Letters,* 3391 (1969).

*The checkers obtained a rotations for *4b*I and *4b*II of $[\alpha]_D$ +1080° and −1082°, respectively.

19. A *cis*(NO_2),*trans*(N)-BIS(*S*-ARGININE)-DINITROCOBALT(III) ISOMER AND ITS USE AS A CATIONIC RESOLVING AGENT

Submitted by M. B. CELAP, M. J. MALINAR, P. N. RADIVOJSA, and Lj. SOLUJIC*
Checked by A. S. KOZLOWSKI,† S. J. ARCHER,‡ and R. D. ARCHER§

Several complex enantiomers have been used for resolving complex ions,[1–3] but the procedures for their preparation are rather lengthy. The preparation of these enantiomers first requires synthesis of the corresponding racemic mixture, which is then treated with an optically active resolving agent to produce two diastereomers. After the separation of diastereomers by fractional crystallization, the removal of the resolving agent affords the corresponding enantiomers, which can be used for the resolution of complex ions. Because of these complications, a new type of resolving agent for complex ions has been proposed, whose preparation is much simpler. This is achieved by using optically active ligands (instead of inactive ones) for the synthesis of the complex, which yields (in only one reaction) two optically active, internal diastereomers as resolving agents.

This has been demonstrated by the synthesis of several diastereomers, including Δ- and Λ-*cis*(NO_2),*trans*(N)-$[Co(NO_2)_2(S\text{-argH})_2]^+$ as the nitrate salts.[4] The analogous Δ isomer as the chloride salt is synthesized herein. The applicability of this product to complex resolutions is shown by the resolution of *cis*(NO_2),*trans*(N)-$[Cogly_2(NO_2)_2]^-$ into diastereomers with the Δ-*cis*(NO_2),*trans*(N)-$[Co(NO_2)_2(S\text{-argH})_2]^+$ ion and by the partial resolution of $[Co(acac)_3]$ on a column of the optically active chloride salt.** The Λ isomer can also be obtained through fractional crystallization, but the temperature dependence precludes the routine synthesis of both isomers without repeated recrystallizations.

A. Δ(−)$_D$-*cis*(NO_2),*trans*(N)-BIS(*S*-ARGININE)-DINITROCOBALT(III) CHLORIDE

$$2\ CoCl_2{\cdot}6H_2O + 4\ S\text{-argH}_2Cl + 4\ NaNO_2 + 2\ NaOH + 1/2\ O_2 = 2\ [Co(NO_2)_2(S\text{-argH})_2]Cl + 6\ NaCl + 15\ H_2O$$

*Chemical Institute, Faculty of Science, University of Belgrade, 11001 Belgrade, Yugoslavia.
†Department of Chemistry, Central Connecticut State College, New Britain, CT 06050.
‡Department of Chemistry, College of William and Mary, Williamsburg, VA 23185.
§Department of Chemistry, University of Massachusetts, Amherst, MA 01003.

**Abbreviations used for ligands: argH = arginine; glyH = glycine; acacH = acetylacetone or 2,4-pentanedione.

Procedure

To a solution of 25.28 g (0.120 mole) of *S*-arginine hydrochloride and 2.40 g (0.060 mole) of NaOH in 80.0 mL of water (in a 250-mL Erlenmeyer flask) the solutions of 14.36 g (0.060 mole) of $CoCl_2 \cdot 6H_2O$ in 40 mL of water and 8.28 g (0.120 mole) of $NaNO_2$ in 40 mL of water are added. Through the resulting solution air is bubbled vigorously for three hours. During this time ~14 g of an orange-brown crystalline *levo*-rotatory (sodium D-line, 589 nm) mixture of diastereoisomers (I) is precipitated, which is filtered on a Buchner funnel and washed successively with ice-cold 48% ethanol, 96% ethanol, and diethyl ether. The filtrate is evaporated on a rotatory evaporator (40–50°) to about a half of its original volume and left at room temperature for three hours, after which precipitate (II) (~5 g) is obtained, filtered, and washed in the same way as in the previous case.*

This operation is repeated once again, but the concentrated filtrate is left to stand overnight. In such a way ~8 g of isomeric mixture (III) is obtained. The early *levo*-rotatory fractions with an [α] of at least −400° are mixed and dissolved in hot (80°) water.** The hot solution is filtered on a Buchner funnel and left at room temperature overnight. During this time ~3.5 g of pure, red-brown crystals of the $\Delta(-)_D$-diastereomer with an $[\alpha]_D^{20} = -550°$† are deposited.‡

By stepwise evaporation of the filtrate, one obtains several additional frac-

*The submitters suggest that (II) is *dextro*-rotatory and that (III) is *levo*. The checkers have found that the composition of the precipitates varies considerably with both the temperature and the rate of bubbling. They recommend a moderate rate of bubbling through a capillary tube, rather than a sintered-glass gas dispersion stick, for a period of 12–18 hours (overnight) at 25° or above. When room temperature is under 25°, the initial precipitate is less *levo*-rotatory or even *dextro*-rotatory. Vigorous bubbling through a gas dispersion tube often results in the premature precipitation of a pasty, impure (low α) precipitate, which overflows the flask and is very difficult to filter. The checkers also recommend only isolation of the $(-)_D$ or Δ isomer for use in further experiments. Early fractions with $[\alpha]_D = -400°$ (or more negative) are most suitable for purification. Sometimes this was the first, sometimes the second, and sometimes both the first and second fractions. The *dextro*-rotatory fractions tend to supersaturate and yield pastes rather than crystalline materials; and the overall yield, when of reasonable rotation, is small. Obviously, the checkers feel that the rotation of each fraction should be measured prior to recrystallization.

**Although the submitters recommended 200 mL of hot water, the checkers used a saturated solution at 80° of ~50 mL for 12 g of complex. About 7 g of pure, red-brown crystals with $[\alpha]_{589} = -550°$ were obtained.

†In aqueous solution (c = 0.1 g/100 mL).

‡The submitters note an ability to obtain appreciable quantities of a *dextro* isomer by its being kept further in a refrigerator. This does not occur if the *levo* fraction is −400° or so. *Dextro* fractions can best be recrystallized using about 40 mL of hot water per 3 g of material. Less water provides pasty precipitates. For samples with $[\alpha]_D$ of +100° or less, the *levo*-rotatory isomer forms first (as orange crystals) when the solution is cooled to room temperature. These can be filtered and the *dextro*-rotatory fraction obtained by cooling in the refrigerator. The *dextro*-rotatory diastereomer appears as a silky yellow precipitate which the checkers found best separated by centrifugation. The submitters obtained sizeable quantities of *dextro*-rotatory material with [α] = +220. The checkers obtained only small amounts (~1 g) with [α] = +250.

tions, which are to be recrystallized until their optical rotations reach the [α] values of the pure isomers. The total yield is ~5 g of Δ-(−)$_D$-*cis*(NO_2),*trans*(N)-[$Co(NO_2)_2$(S-argH)$_2$]Cl·1/2H_2O with $[\alpha]_D^{20}$ = −550°. For analysis the substance is dried at 105° for two hours. During this time the mass of the (−)$_D$-diastereomer decreases by 1.45%, which corresponds to half of a water molecule per mole (1.65%). *Anal.* Calcd. for $CoN_{10}C_{12}H_{28}O_8Cl$ (MW = 534.62): C, 26.96; N, 26.20; H, 5.28. Found for the Δ-(−)$_D$-diastereomer: C, 26.87; N, 26.31; H, 5.16.*

Properties

Both diastereomers are stable and inert substances, even in aqueous solution at higher temperatures (inversion half-life at 65° for the Λ-(+)$_D$-diastereomer is 356 ± 20 hours, and for the Δ-(−)$_D$-diastereomer, it is 172 ± 9 hours).[5] Because of this stability, it is possible to use the same sample of a diastereomer for several resolutions of complex racemic mixtures, after its recovery by elution with 1 *M* NaCl from the ionic exchanger used in obtaining the enantiomers (see below).

B. RESOLUTION OF SILVER *cis*(NO_2),*trans*(N)-DIGLYCINATO-DINITROCOBALTATE(III)†

2 Δ(−)$_D$-*cis*(NO_2),*trans*(N)-[$Co(NO_2)_2$(S-argH)$_2$]Cl·1/2H_2O +
2 Ag{*cis*(NO_2),*trans*(N)-[Cogly$_2$(NO_2)$_2$]} + 6 H_2O =
(−)$_D${Δ-(−)$_D$-*cis*(NO_2),*trans*(N)-[$Co(NO_2)_2$(S-argH)$_2$]}-
{Λ(+)$_D$-*cis*(NO_2),*trans*(N)-[Cogly$_2$(NO_2)$_2$]}·2H_2O +
(−)$_D${Δ-(−)$_D$-*cis*(NO)$_2$),*trans*(N)-[$Co(NO_2)_2$(S-argH)$_2$]}-
{Δ(−)$_D$-*cis*(NO_2),*trans*(N)-[Cogly$_2$(NO_2)$_2$]}·5H_2O + 2 AgCl

Procedure

To a hot (70°) solution of 5.35 g (0.01 mol) of Δ-(−)$_D$-*cis*(NO_2),*trans*(N)[$Co(NO_2)_2$](S-argH)$_2$]Cl·1/2H_2O in 200 mL of water, a suspension of 4.06 g (0.01 mol) of *cis*(NO_2),*trans*(N)-Ag[Cogly$_2$(NO_2)$_2$][7] in 30 mL of water (70°), is added. The solution mixture is stirred at 70° for 20 minutes, and then it is left to cool at room temperature. The precipitated silver chloride

*The *levo*-rotatory or delta diastereomer was dried by the checkers for two hours at 100° and analyzed as a monohydrate: *Anal.* Calcd. for $CoN_{10}C_{12}H_{30}O_9Cl$ (MW = 552.6); C, 26.06; N, 25.33; H, 5.42. Found: C, 26.13; N, 25.27; H, 5.70.

†The resolution of the corresponding potassium salt by means of *d*-cinchonine has already been described by Celap and co-workers.[6]

is separated by filtration and the filtrate is concentrated on a rotatory vacuum evaporator at 60°* until the appearance of crystals, and then it is left to stand at room temperature. After two hours, 3.2 g of pure $(-)_D$-[Δ-$(-)_D$-*cis*(NO_2), *trans*(N)-Co(NO_2)$_2$-(S-argH)$_2$][Λ-$(+)_D$-*cis*(NO_2),*trans*(N)-[Cogly$_2$-(NO_2)$_2$]·$2H_2O$ (I), is separated from the solution in the form of light-brown, crystalline plates with $[\alpha]_D^{20} = -175°$ (c = 0.1 g/100 mL of water).† The filtrate obtained after the separation of diastereomer I is left to stand overnight in a refrigerator, whereupon 3.4 g of the second diastereomer pentahydrate (II) crystallizes in the form of yellow-orange needles. The substance is recrystallized from hot (70°) water until a constant optical rotation $[\alpha]_D^{20} = -515°$ (c = 0.1 g/100 mL water) is obtained, yielding 3.1 g of pure diastereomer II.‡

For analysis, the isolated substances I and II are dried at 105° for two hours, whereupon I loses 4.14% of its mass, which corresponds to a loss of two water molecules (3.93%), whereas diastereomer II loses 10.35%, which corresponds to a loss of five water molecules (10.13%). *Anal.* Calcd. for $Co_2C_{16}N_{14}O_{16}H_{36}$ (MW = 798.07): C, 24.06; N, 24.56; H, 4.47. Found for diastereomer I: C, 24.48; N, 24.78; H, 4.69; and for diastereomer II: C, 23.86; N, 24.63; H, 4.91.

A solution of 3 g of I or II in 50 mL of water is passed through a cationic Dowex 50W X4 column (25 cm long, 1.5 cm in diameter), in the potassium form, and the column is eluted with water.§ Each eluate obtained is concentrated on a rotatory vacuum evaporator at 70° to half of its original volume, to which 0.1 *M* silver nitrate solution is added, until no additional precipitate is separated. The precipitates obtained are filtered on a Buchner funnel, successively washed with water, 96% ethanol, and diethyl ether, giving the *cis*(NO_2),*trans*(N)-Ag[Cogly$_2$(NO_2)$_2$] enantiomers in almost quantitative yield. The optical rotation of the isolated enantiomers is $[\alpha]_D^{20} = \pm 370°$ (c = 0.1 g/100 mL water.)** *Anal.* Calcd. for $AgCoC_4H_8O_8N_4$ (MW = 406.95): C, 11.80; N, 13.77; H, 1.98. Found for the Λ-$(+)_D$-enantiomer: C, 12.04; N, 13.88; H, 2.23; and for the Δ-$(-)_D$-enantiomer: C, 12.10; N, 14.05; H, 1.99.

*The filtrate should be concentrated at 60° down to 100 mL and allowed to cool according to the checkers. The solution has a tendency to supersaturate and sometimes crystals don't form until the volume is much lower, in which case the crystals are a mixture of both diastereomers.

†The checkers washed the crystals with 48% ethanol, 96% ethanol, and diethyl ether. In some cases the diastereomer needed to be recrystallized. $[\alpha]_D^{20} = -185°$.

‡The checkers obtained $[\alpha]_D^{20} = -519°$. Any brown prisms of diastereomer I which are visible (as diastereomer II dissolves) must be removed by either filtration or decantation, in order to obtain pure diastereomer II. The checkers washed the crystals with 48% ethanol, 96% ethanol, and diethyl either. A yield of 2 g (after recrystallization) was obtained.

§The checkers used Dowex 50W X8 and found that the sodium form can also be employed. However, with either form, the column is exhausted after one use. Only a small fraction of the complex cation can be removed from the more highly crosslinked resin, even with saturated NaCl solutions.

**The checkers obtained optical rotations of −360° and +370°.

C. PARTIAL RESOLUTION OF TRIS(2,4-PENTANEDIONATO)COBALT(III) BY COLUMN CHROMATOGRAPHY WITH Δ-(−)$_D$-*cis*(NO_2),*trans*(N)-[$Co(NO_2)_2$(S-argH)$_2$]Cl·H_2O[8]

Procedure

Approximately 25 g of finely ground Δ-(−)$_D$-[$Co(NO_2)_2$(S-argH)$_2$]Cl is sieved with a 140-mesh screen, and the fines are discarded. The ~20 g of complex which does not pass through the screen are slurried in toluene and packed into a chromatography column 1 cm × 50 cm. A 0.2-g sample of tris(2,4-pentanedionato)cobalt(III)[9] dissolved in 5 mL of toluene is applied to the top of the column. The column is eluted with toluene and 35 mL of eluate is collected prior to the emergence of the sample band. Sixteen fractions of 0.8 mL each (15 drops) are collected and each is diluted to 3.0 mL. From the absorbance of the solutions at 590 nm (ϵ = 160 L mole^{-1}cm^{-1}) and the optical rotations at 546 nm, molar rotations are obtained for the earliest and latest fractions collected. The middle fractions are almost inactive. There is a large error at the last two concentrations.

Fraction	Absorbance	$[M]^{20°}_{546}$
1	0.61	−250
2	1.03	−280
...	...	...
14	0.12	+380
15	0.04	+700
16	0.03	+300

An equal (or a little higher) degree of resolution by chromatography on D-lactose has been reported,[10,11] but this entails the use of much longer columns and much slower elution rates. The resolution obtained is about 1% of the molar rotation of ±30,000 estimated for complete resolution.[12]

References

1. S. Kirschner, in *Preparative Inorganic Reactions,* Vol. 1, W. L. Jolly (ed.), Interscience Publishers, New York, 1964, p. 29.
2. A. M. Sargeson, in *Chelating Agents and Metal Chelates,* F. T. Dwyer and D. T. Mellor (eds.), Academic Press, New York, 1965, p. 183.
3. W. T. Jordan and L. R. Froebe, *Inorg. Syn.,* **18,** 96 (1978).
4. (a) M. B. Celap, B. A. Kamberi, and T. J. Janjic, *Bull. Soc. Chim. Beograd,* **35,** 158 (1970) (in English); (b) W. H. Watson, D. R. Johnson, M. B. Celap, and B. Kamberi, *Inorg. Chim. Acta,* **6,** 591 (1972); (c) B. A. Kamberi, M. B. Celap, and T. J. Janjic, *Bull. Soc. Chim. Beograd,* **43,** 149 (1978) (in English).

5. M. B. Celap, L. J. Aljancic, and T. J. Janjic, *J. Indian Chem. Soc.*, **59,** 1255 (1982).
6. M. B. Celap, D. J. Radanovic, and T. J. Janjic, *Inorg. Chem.*, **4,** 1494 (1965).
7. M. B. Celap, T. J. Janjic, and D. J. Radanovic, *Inorg. Syn.*, **9,** 173 (1967).
8. M. B. Celap, I. M. Hodzic, and T. J. Janjic, *J. Chromatog.*, **198,** 172 (1980).
9. B. E. Bryant and W. C. Fernelius, *Inorg. Syn.*, **5,** 188 (1950).
10. J. P. Collman, R. P. Blair, R. L. Marshall, and S. Slade, *Inorg. Chem.*, **2,** 576 (1963).
11. R. C. Fay, A. Y. Girgis, and U. Klabunde, *J. Am. Chem. Soc.*, **92,** 7056 (1970).
12. J.-Y. Sun, Ph.D. Dissertation, University of North Carolina, 1967; *Dissert. Abstr., B,* **28,** 4482 (1968).

20. RACEMIC AND OPTICALLY ACTIVE COBALT(III) COMPLEXES OF CDTA,* EDTA,* AND PDTA* IN NON-AQUEOUS SOLVENTS

Submitted by L. H. O'CONNOR and K. H. PEARSON†
Checked by H. HEASTER and P. HOGGARD‡

Dwyer and Garvan[1] prepared racemic K[Co(cdta)]·$3H_2O$ and K[Co(pdta)]·H_2O, in water solution using racemic cdta[2] and pdta.[1] They also prepared enantiomers of the complex anions by using the optical enantiomers of *cis*-$[Co(en)_2(NO_2)_2]Br$ as the resolving agents.[3] Zadnik and Pearson[4] prepared Cs[Co(R,R(−)cdta)] using previously synthesized R,R(−)H_4cdta.[5]

The syntheses of $[Co(edta)]^-$ complexes historically have involved aqueous systems requiring many hours of preparation, large volumes of solvent, and unwanted side reactions. Brintzinger, Thiele, and Müller[6] obtained the Na[Co(edta)] complex as a 4-hydrate and found that the compound could be dehydrated by drying at 150° with no change in its properties. Schwarzenbach[7] synthesized $Ba[Co(edta)]_2 \cdot 4H_2O$ in a similar manner using bromine as the oxidizing agent. Kirschner[8] and Dwyer and co-workers[9] also prepared this complex in aqueous solution. The racemic $[Co(edta)]^-$ complex was resolved by Busch and Bailar,[10] Dwyer and Garvan,[11] and Jordan and Froebe[12] using various optically active resolving agents.

An essentially nonaqueous procedure is described for preparing the racemic and optically active forms of the heavy alkali metal complexes (potassium, rubidium, and cesium) of the cobalt(III) complexes of cdta, edta, and pdta. The

*cdta = *trans*-[N,N′-1,2-cyclohexanediylbis[N-(carboxymethyl)glycinate]]$^{4-}$ (*trans*-1,2-cyclohexanediaminetetraacetate^{4-});
edta = [N,N′-1,2-ethanediylbis[N-(carboxymethyl)glycinate]]$^{4-}$ (ethylenediaminetetraacetate^{4-});
pdta = [N,N′-(1-methyl-1,2-ethanediyl)bis[N-(carboxymethyl)glycinate]]$^{4-}$ (propylenediaminetetraacetate^{4-}).

†Department of Chemistry, Cleveland State University, Cleveland, OH 44115.
‡Department of Chemistry, North Dakota State University, Fargo, ND 58105.

compounds can be synthesized in two to three hours, with 81–89% yields and high purity.

A. CESIUM *trans*-[*N*,*N*′-1,2-CYCLOHEXANEDIYLBIS[*N*-(CARBOXYMETHYL)GLYCINATO]]COBALTATE(III) AND ITS ENANTIOMERS

1. $Cs[Co(cdta)]\cdot H_2O$

$$CoCO_3 + C_{14}H_{22}O_8N_2 \longrightarrow H_2[Co(C_{14}H_{18}N_2O_8)] + CO_2 + H_2O$$

$$2H_2[Co(C_{14}H_{18}N_2O_8)] + H_2O_2 \longrightarrow 2H[Co(C_{14}H_{18}N_2O_8)] + 2H_2O$$

$$H[Co(C_{14}H_{18}N_2O_8)] + CsOH \longrightarrow Cs[Co(C_{14}H_{18}N_2O_8)]\cdot H_2O$$

Procedure

A dry mixture of analytical reagent grade cobalt(II) carbonate (0.595 g, 0.005 mole) and racemic H_4cdta (1.73 g, 0.005 mole) are placed into a 50-mL Erlenmeyer flask and 4.0 mL of deionized water is added. The slurry is heated gradually to 70° with constant stirring on a magnetic stirring hot plate. When the evolution of carbon dioxide ceases, 1.5 mL of fresh 30% hydrogen peroxide is added dropwise to oxidize the resulting Co(II) complex. The mixture is stirred at 70° for 15 minutes, the reaction flask is removed from the hot plate, and cesium hydroxide (0.750 g, 0.005 mole) is added slowly. The mixture is allowed to react for an additional 15 minutes and is then cooled.

Note. All weighings and subsequent steps are to be performed in a nitrogen-purged drybox. The warm mixture is filtered through a medium porosity glass-fritted crucible into a 400-mL beaker. The reaction flask is washed with three 2-mL aliquots of ethanol and these are put through the filter. Cold absolute ethanol (300 mL) is added slowly and with continuous stirring. The ethanolic filtrate is kept cold (0°) and stirred vigorously for 30 minutes. The $Cs[Co(cdta)]\cdot H_2O$ is removed by filtration, first by aspiration through a Büchner funnel (S + S filter #589 white) and then through a medium glass-fritted crucible. The precipitates are washed with absolute ethanol and acetone, and then dried in a vacuum oven at 120° for 16 hours. The potassium complex, $K[Co(cdta)]\cdot 1H_2O$, and the rubidium complex, $Rb[Co(cdta)]\cdot 1H_2O$, are synthesized in an analogous manner, except that 0.281 g (0.005 mole) of potassium hydroxide or 0.512 g (0.005 mole) of rubidium hydroxide is substituted for cesium hydroxide.

2.a. $\Delta(+)_{546}$-Cs[Co(R,R(−)cdta)]·$1H_2O$

The same procedure used for the preparation of racemic $Cs[Co(cdta)]\cdot 1H_2O$ is followed, except that R,R(−)H_4cdta replaces racemic H_4cdta. *Anal.* Calcd. for

Cs[Co(R,R(−)cdta)]·1H_2O: C, 30.45; H, 3.65; N, 5.07. Found: C, 30.21; H, 3.92; N, 5.03.

2.b. $\Delta(+)_{546}$-Rb[Co(R,R(−)cdta)]·1H_2O

The same procedure as that given for the preparation of Cs[Co(cdta)]·1H_2O is followed, except that R,R(−)H_4cdta replaces racemic H_4cdta, and rubidium hydroxide (0.512 g, 0.005 mole) is substituted for cesium hydroxide. *Anal.* Calcd. for Rb[Co(R,R(−)cdta)]·1H_2O: C, 33.32; H, 3.99; N, 5.55. Found: C, 33.33; H, 4.25; N, 5.51.

2.c. $\Delta(+)_{546}$-K[Co(R,R(−)cdta)]·1H_2O

The same procedure as that given for the preparation of Cs[Co(cdta)]·1H_2O is followed, except that R,R(−)H_4cdta replaces racemic H_4cdta, and potassium hydroxide (0.281 g, 0.005 mole) is substituted for cesium hydroxide. *Anal.* Calcd. for K[Co(R,R(−)cdta)]·1H_2O: C, 36.69; H, 4.40; N, 6.11. Found: C, 37.05; H, 4.94; N, 6.41.

Properties

The absorption data and yields for the six complexes are reported in Table I. The specific and molar rotations at 546 nm and 365 nm are given in Table II. The optical rotatory dispersion and circular dichroism properties determined in aqueous solutions are given in Tables III and IV. The racemic complexes were calculated as monohydrate compounds, since the elemental analyses for the optically active complexes correspond to monohydrate compounds, and both the optically active and racemic complexes give similar absorption data. The ORD and CD spectra were obtained on modified computerized Perkin Elmer Model 241 and Cary Model 61 Spectropolarimeters, respectively.[13,14]

B. CESIUM *trans*-[*N*,*N*′-1,2-ETHANEDIYLBIS[*N*-(CARBOXYMETHYL)GLYCINATO]]COBALTATE(III) AND ITS ENANTIOMERS

1. Cs[Co(edta)]·2H_2O

$$CoCO_3 + C_{10}H_{16}N_2O_8 \longrightarrow H_2[Co(C_{10}H_{12}N_2O_8)] + CO_2 + H_2O$$

$$2H_2[Co(C_{10}H_{12}N_2O_8)] + H_2O_2 + 2CsOH \longrightarrow 2Cs[Co(C_{10}H_{12}N_2O_8)]\cdot 2H_2O$$

Procedure

A dry mixture of reagent grade cobalt(II) carbonate (0.595 g, 0.005 mole) and previously recrystallized H_4edta (1.46 g, 0.005 mole) are placed into a 50-mL

TABLE I Absorption Data and Yields

Compounds ($\cdot 1H_2O$)	Yields (%)	Concentration ($M \times 10^3$)	540 nm		383 nm		227 nm[b]	
			A	ϵ[a]	A	ϵ	A	ϵ
K[Co(cdta)]	88.7	2.788	0.856	307	0.573	206	0.579	20,800
Δ-$(+)_{546}$-K[Co(R,R(−)cdta)]	87.4	1.601	0.478	299	0.345	216	0.332	20,700
Rb[Co(cdta)]	90.9	1.965	0.593	302	0.409	208	0.412	21,000
Δ-$(+)_{546}$-Rb[Co(R,R(−)cdta)]	89.8	2.001	0.609	304	0.415	207	0.414	20,700
Cs[Co(cdta)]	88.5	1.886	0.581	308	0.398	211	0.395	20,900
Δ-$(+)_{546}$-Cs[Co(R,R(−)cdta)]	89.9	2.710	0.820	303	0.562	207	0.559	20,600

[a] ϵ = 1 mole^{-1} cm^{-1}.

[b] Diluted 1:100, Molarity $\times 10^5$.

TABLE II Specific and Molar Rotations[a]

Compounds ($\cdot 1H_2O$)	$[\alpha]_{546}$	$[M]_{546}$	$[\alpha]_{365}$	$[M]_{365}$
Δ-$(+)_{546}$-K[Co(R,R(−)cdta)]	+1090	+5000	−732	−3360
Δ-$(+)_{546}$-Rb[Co(R,R(−)cdta)]	+1000	+5050	−673	−3400
Δ-$(+)_{546}$-Cs[Co(R,R(−)cdta)]	+903	+4990	−602	−3320

[a] $[\alpha]$ = deg·mL·g^{-1}dm^{-1}. $[M]$ = deg·cm^{2}dmole^{-1} = $[\alpha]$ $M/100$.

Erlenmeyer flask, and 4.0 mL of deionized water is added. The mixture is then treated exactly as the cdta complex in Section A-1.

Anal. Calcd. for Cs[Co(edta)]·$2H_2O$: C, 23.27; H, 3.12; N, 5.43. Found: C, 23.55; H, 3.21; M, 5.50.

2. K[Co(edta)]·$2H_2O$

The procedure given for the preparation of Cs[Co(edta)]·$2H_2O$ is followed, except that 0.281 g (0.005 mole) of potassium hydroxide is substituted for the cesium hydroxide. The filtration step for this complex is slow (35 minutes), in comparison with the cesium and rubidium complexes.

Anal. Calcd. for K[Co(edta)]·$2H_2O$: C, 28.44; H, 3.82; N, 6.63. Found: C, 28.23; H, 3.70; N, 6.56.

3. Rb[Co(edta)]·$2H_2O$

The procedure given for the preparation of Cs[Co(edta)]·$2H_2O$ is followed, except that 0.512 g (0.005 mole) of rubidium hydroxide is substituted for cesium hydroxide. Based on analogous absorption values, the rubidium complex is calculated to contain two molecules of water of hydration.

Properties

The products, transparent platelets, are deep rose by transmitted light and violet by reflected light. The absorption data and yields for the three complexes are given in Table V.

C. CESIUM [*N,N′*-(1-METHYL-1,2-ETHANEDIYL)BIS[*N*-(CARBOXYMETHYL)GLYCINATO]]COBALTATE(III) AND ITS ENANTIOMERS

1. Cs[Co(pdta)]·H_2O

$$CoCO_3 + C_{11}H_{18}N_2O_8 \longrightarrow H_2[Co(C_{11}H_{14}N_2O_8)] + CO_2 + H_2O$$

$$2\ H_2[Co(C_{11}H_{14}N_2O_8)] + H_2O_2 \longrightarrow 2\ H[Co(C_{11}H_{14}N_2O_8)] + 2\ H_2O$$

$$H[Co(C_{11}H_{14}N_2O_8)] + CsOH \longrightarrow Cs[Co(C_{11}H_{14}N_2O_8)]\cdot H_2O$$

TABLE III ORD Peak and Trough Data

Compounds	$[M]$ (deg·cm^{2}dmole^{-1})					
(·1H_2O)	562 nm	462 nm	404 nm	376 nm	352 nm	262 nm
Δ-(+)$_{546}$-K[Co(R,R(−)cdta)]	+6410	−4050	−2510	−3670	−2900	−23,200
Δ-(+)$_{546}$-Rb[Co(R,R(−)cdta)]	+6400	−4000	−2550	−3670	−2980	−23,300
Δ-(+)$_{546}$-Cs[Co(R,R(−)cdta)]	+6390	−3930	−2550	−3680	−2990	−23,300

TABLE IV CD Maxima Data

Compounds	593 nm		527 nm		240 nm	
(·1H_2O)	[θ][a]	Δϵ[b]	[θ]	Δϵ	[θ]	Δϵ
Δ-(+)$_{546}$-K[Co(R,R(−)cdta)]	−5370	−1.63	+6190	+1.88	−52,000	−15.8
Δ-(+)$_{546}$-Rb[Co(R,R(−)cdta)]	−5320	−1.61	+6260	+1.90	−50,900	−15.4
Δ-(+)$_{546}$-Cs[Co(R,R(−)cdta)]	−5360	−1.62	+6300	+1.91	−51,400	−15.6

[a] [θ] = deg·cm^{2}dmole^{-1}.
[b] Δϵ = l mole^{-1} cm^{-1}.

TABLE V Absorption Data and Yields[a]

Complex	Yield (%)	Concentration (molarity)	λ_{max} (nm)	A	ϵ_{max} l $mole^{-1}$ cm^{-1}
$K[Co(edta)]\cdot 2H_2O$	87.3	2.761×10^{-3}	540	0.907	328
		2.761×10^{-3}	383	0.615	223
		2.761×10^{-5}	227	0.607	22,000
$Rb[Co(edta)]\cdot 2H_2O$	86.4	3.790×10^{-3}	540	1.24	327
		3.790×10^{-3}	383	0.841	222
		3.790×10^{-5}	227	0.829	21,900
$Cs[Co(edta)]\cdot 2H_2O$	88.0	2.976×10^{-3}	540	0.955	322
		2.976×10^{-3}	383	0.648	221
		2.976×10^{-5}	227	0.634	21,700

[a]Literature,[15] $\lambda = 537$ nm ($\epsilon = 324$), 384 (229), 226 (21,800). Literature,[16] $\lambda = 538$ nm ($\epsilon = 347$), 375 (246). Literature,[11] *l*-$K[Co(edta)]\cdot 2H_2O$: $[\alpha]_D = +150°$; $[\alpha]_{546} = -1000°$. Values obtained from material synthesized by this method: $[\alpha]_D = +160°$, $[\alpha]_{546} = -1010°$. (Using instrument described in Reference 13). $\lambda = 540$ nm ($\epsilon = 328$), 383 (222), 227 (22,400). Literature,[12] $K(-)_{546}$-$[Co(edta)]\cdot 2H_2O$: $[\alpha]_{546} = -1000°$.

TABLE VI UV-VIS Absorption Data

Compounds	Yield	Concentration	540 nm		383 nm		227 nm[a]	
($\cdot 1H_2O$)	(%)	(molarity × 10^3)	A	ε	A	ε	A	ε
K[Co(pdta)]	86.7	2.984	0.853	286	0.590	198	0.595	20,000
Δ-(+)$_{546}$-K[Co(R(−)pdta)]	82.4	2.878	0.827	287	0.587	204	0.590	20,500
Λ-(+)$_{546}$-K[Co(S(+)pdta)]	81.0	2.462	0.728	296	0.511	208	0.514	20,900
Rb[Co(pdta)]	88.0	2.712	0.780	287	0.538	199	0.549	20,300
Δ-(+)$_{546}$-Rb[Co(R(−)pdta)]	84.6	2.239	0.653	291	0.465	208	0.453	20,300
Λ-(+)$_{546}$-Rb[Co(S(+)pdta)]	82.5	2.231	0.665	299	0.459	206	0.473	21,200
Cs[Co(pdta)]	89.1	2.472	0.734	296	0.505	204	0.509	20,600
Δ-(+)$_{546}$-Cs[Co(R(−)pdta)]	88.0	2.289	0.669	292	0.463	202	0.472	20,600
Λ-(+)$_{546}$-Cs[Co(S(+)pdta)]	84.8	1.727	0.497	288	0.345	200	0.359	20,800

[a]Diluted 1:100, Molarity × 10^5. $\epsilon = 1\ mole^{-1}\ cm^{-1}$.

Procedure

A dry mixture of analytical reagent grade cobalt(II) carbonate (0.595 g, 0.005 mole) and previously recrystallized H_4pdta (1.53 g, 0.005 mole) are placed into a 50-mL Erlenmeyer flask, and 6.0 mL of deionized H_2O is added. The mixture is then treated exactly as the cdta complex in Section A-1. *Anal.* Calcd. for Cs[Co(pdta)]·1H_2O: C, 25.80; H, 3.15; N, 5.47. Found: C, 25.90; H, 3.44; N, 5.40. The potassium and rubidium complexes are synthesized in a similar manner, except that potassium hydroxide (0.281 g, 0.005 mole) and rubidium hydroxide (0.512 g, 0.005 mole) are substituted for cesium hydroxide, respectively.

2.a. Δ(+)$_{546}$-Cs[Co(R(−)pdta)]·1H_2O

The procedure for the preparation of racemic Cs[Co(pdta)]·H_2O is followed, except that R(−)H_4pdta[1] replaces racemic H_4pdta, and only 4.0 mL, instead of 6.0 mL, of water is used in the synthesis. *Anal.* Calcd. for Cs[Co(R(−)pdta)]·1H_2O: C, 25.80; H, 3.15; N, 5.47. Found: C, 25.62; H, 3.42; N, 5.47.

2.b. Δ(+)$_{546}$-Rb[Co(R(−)pdta)]·1H_2O

The procedure given for the preparation of Cs[Co(pdta)]·1H_2O is followed, except that R(−)H_4pdta replaces H_4pdta, only 4.0 mL, instead of 6.0 mL, of water is used, and rubidium hydroxide (0.512 g, 0.005 mole) is substituted for cesium hydroxide. *Anal.* Calcd. for Rb[Co(R(−)pdta)]·1H_2O: C, 28.43; H, 3.47; N, 6.03. Found: C, 28.70; H, 3.58; N, 5.87.

2.c. Δ(+)$_{546}$-K[Co(R(−)pdta)]·1H_2O

The procedure given for the preparation of Cs[Co(pdta)]·1H_2O is followed, except that R(−)H_4pdta replaces H_4pdta, only 4.0 mL of water is used, and potassium hydroxide (0.281 g, 0.005 mole) is substituted for cesium hydroxide.

TABLE VII Specific and Molar Rotations[a]

Compounds (·1H_2O)	$[\alpha]_{546}$	$[M]_{546}$	$[\alpha]_{365}$	$[M]_{365}$
Δ-(+)$_{546}$-K[Co(R(−)pdta)]	+880	+3680	−586	−2450
Δ-(−)$_{546}$-K[Co(S(+)pdta)]	−883	−3690	+587	+2460
Δ-(+)$_{546}$-Rb[Co(R(−)pdta)]	+787	+3660	−518	−2410
Δ-(−)$_{546}$-Rb[Co(S(+)pdta)]	−795	−3690	+536	+2490
Δ-(+)$_{546}$-Cs[Co(R(−)pdta)]	+716	+3670	−494	−2530
Δ-(−)$_{546}$-Cs[Co(S(+)pdta)]	−711	−3640	+495	+2530

[a] $[\alpha]$ = deg·mL $g^{-1}dm^{-1}$. $[M] = [\alpha]\,M/100$ (deg·cm^2dmole^{-1}).

TABLE VIII ORD Peak and Trough Data

Compounds ($\cdot 1H_2O$)	[*M*] deg·cm²dmole⁻¹					
	554 nm	443 nm	404 nm	374 nm	347 nm	260 nm
Δ-$(+)_{546}$-K[Co(R(−)pdta)]	+3880	−2330	−1420	−2650	−2130	−16,700
Λ-$(-)_{546}$-K[Co(S(+)pdta)]	−3960	+2370	+1360	+2640	+2090	+17,000
Δ-$(+)_{546}$-Rb[Co(R(−)pdta)]	+3780	−2250	−1370	−2580	−2050	−17,200
Λ-$(-)_{546}$-Rb[Co(S(+)pdta)]	−3840	+2320	+1370	+2630	+2130	+17,100
Δ-$(+)_{546}$-Cs[Co(R(−)pdta)]	+3930	−2360	−1420	−2740	−2230	−17,000
Λ-$(-)_{546}$-Cs[Co(S(+)pdta)]	−3890	+2290	+1430	+2710	+2220	+17,300

TABLE IX CD Maxima Data[a]

Compounds ($\cdot 1H_2O$)	586 nm		513 nm		237 nm	
	[θ]	Δϵ	[θ]	Δϵ	[θ]	Δϵ
Δ-$(+)_{546}$-K[Co(R(−)pdta)]	−4610	−1.40	+2540	+0.770	−25,000	−7.58
Λ-$(-)_{546}$-K[Co(S(+)pdta)]	+4790	+1.45	−2650	−0.803	+25,600	+7.76
Δ-$(+)_{546}$-Rb[Co(R(−)pdta)]	−4580	−1.39	+2480	+0.752	−24,800	−7.52
Λ-$(-)_{546}$-Rb[Co(S(+)pdta)]	+4760	+1.44	−2520	−0.764	+25,300	+7.67
Δ-$(+)_{546}$-Cs[Co(R(−)pdta)]	−4840	−1.47	+2540	+0.770	−26,700	−8.09
Λ-$(-)_{546}$-Cs[Co(S(+)pdta)]	+4810	+1.42	−2520	−0.763	+26,000	+7.89

[a] $[\theta]$ = deg·cm²dmole⁻¹. $\Delta\epsilon$ = 1 mole⁻¹cm⁻¹.

Anal. Calcd. for K[Co(R(−)pdta)]·1H_2O: C, 31.59; H, 3.86; N, 6.70. Found: C, 31.90; H, 4.24; N, 6.80.

3. $\Lambda(-)_{546}$-Cs[Co(S(+)pdta)]·1H_2O

The procedure given for the preparation of Cs[Co(pdta)]·1H_2O is followed, except that S(+)H_4pdta,[17] replaces H_4pdta and only 4.0 mL, instead of 6.0 mL, of water is used in the synthesis. The potassium complex Λ-$(-)_{546}$-K[Co-(S(+)pdta)]·1H_2O, and the rubidium complex Λ-$(-)_{546}$-Rb[Co(S(+)pdta)]·1H_2O are synthesized in an analogous manner, except that 0.281 g (0.005 mole) of potassium hydroxide and 0.512 g (0.005 mole) of rubidium hydroxide, respectively, are substituted for cesium hydroxide.

Properties

The absorption data and yields for the nine complexes are reported in Table VI. The specific and molar rotations at 546 nm and 365 nm are given in Table VII. The optical rotatory dispersion and circular dichroism spectra (in aqueous solution) are given in Tables VIII and IX. The ORD and CD spectra were obtained on modified computerized Perkin Elmer Model 241 and Cary Model 61 Spectropolarimeters, respectively.[13,14]

References

1. (a) F. P. Dwyer and F. L. Garvan, *J. Am. Chem. Soc.*, **83,** 2610 (1961); (b) *Ibid.*, **81,** 2955 (1955).
2. G. Schwarzenbach and H. Ackerman, *Helv. Chim. Acta,* **32,** 1682 (1949).
3. F. P. Dwyer and F. L. Garvan, *Inorg. Synth.*, **6,** 195 (1960).
4. V. C. Zadnik and K. H. Pearson, *Anal. Lett.*, **12,** 1267 (1979).
5. P. E. Reinbold and K. H. Pearson, *Inorg. Chem.*, **9,** 2325 (1970).
6. H. Brintzinger, H. Thiele, and U. Muller, *Z. anorg. Chem.*, **251,** 285 (1943).
7. G. Schwarzenbach, *Helv. Chim. Acta,* **32,** 839 (1949).
8. S. Kirschner, *Inorg. Synth.*, **5,** 196 (1957).
9. F. P. Dwyer, E. C. Gyarfas, and D. P. Mellor, *J. Phys. Chem.*, **59,** 296 (1955).
10. D. H. Busch and J. C. Bailar, Jr., *J. Am. Chem. Soc.*, **75,** 4574 (1953).
11. F. P. Dwyer and F. L. Garvan, *Inorg. Synth.*, **6,** 192 (1960).
12. W. T. Jordan and L. R. Froebe, *Inorg. Synth.*, **18,** 100 (1979).
13. V. C. Zadnik, J. L. Scott, R. Megargle, J. Kerkay, and K. H. Pearson, *J. Automatic Chem.*, **4,** 206 (1979).
14. V. C. Zadnik, K. H. Pearson, and R. Megargle, *Ibid.*, **3,** 138 (1981).
15. J. Hidaka, Y. Shimura, and R. Tsuchida, *Bull. Chem. Soc., Japan,* **35,** 567 (1962).
16. B. E. Douglas, R. A. Haines, and J. G. Brushmiller, *Inorg. Chem.*, **2,** 1194 (1963).
17. P. E. Reinhold, Ph.D. Dissertation, Texas A & M University (1970).

Chapter Four

BRIDGE AND CLUSTER COMPOUNDS

21. μ-CARBOXYLATODI-μ-HYDROXO-BIS[TRIAMMINECOBALT(III)] COMPLEXES

Submitted by K. WIEGHARDT* and H. SIEBERT†
Checked by E. S. GOULD‡ and V. S. SRINIVASAN‡

The preparation in low yields of sulfate, chloride, bromide, nitrate, and dithionate salts of the binuclear cation tri-μ-hydroxo-bis[triamminecobalt(III)] has been reported by A. Werner.[1] An X-ray analysis[2] of the dithionate salt, $[(NH_3)_3Co(OH)_3Co(NH_3)_3]_2(S_2O_6)_3$, confirmed the structure of the cation to consist of two face-sharing octahedra around cobalt(III) connected *via* three hydroxo-bridges. The Co—Co distance is relatively short (2.565 Å).

An improved synthesis yielding the perchlorate salt, $[(NH_3)_3Co(OH)_3Co(NH_3)_3](ClO_4)_3 \cdot 2\,H_2O$ has been reported.[3] We report here an improved synthesis[4] for *trans*-$[Co(NH_3)_3(NO_2)_3]$, originally described by Jörgensen.[5]

The chemistry of the tri-μ-hydroxo-bis[triamminecobalt(III)]$^{3+}$ cation has attracted much interest. The perchlorate salt is the starting material for many new complexes—mononuclear complexes of the type $[Co(NH_3)_3X_3]$ and binuclear

*Lehrstuhl für Anorganische Chemie I der Ruhr-Universität Bochum, D-4630 Bochum 1, Federal Republic of Germany.

†Anorganisch-Chemisches Institut der Universität Heidelberg, D-6900 Heidelberg, Federal Republic of Germany.

‡Department of Chemistry, Kent State University, Kent, OH 44242.

complexes of the type $[(NH_3)_3LCo(OH)_2CoL(NH_3)_3]^{2+}$ [3] as well as bridged binuclear complexes of the types

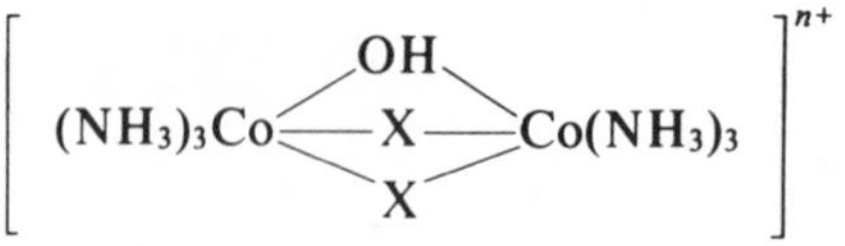

$$\left[(NH_3)_3Co \begin{matrix} OH \\ -X- \\ X \end{matrix} Co(NH_3)_3 \right]^{n+}$$

and

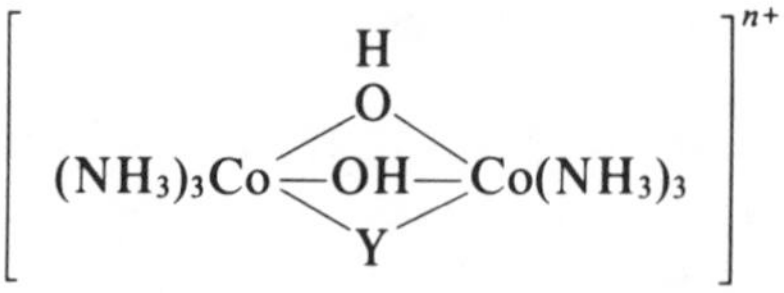

$$\left[(NH_3)_3Co \begin{matrix} H \\ O \\ -OH- \\ Y \end{matrix} Co(NH_3)_3 \right]^{n+}$$

where

L = F^-, NCS^-, N_3^-, H_2O [3] = external anion
X = $CH_2ClCO_2^-$, $CHCl_2CO_2^-$, $CCl_3CO_2^-$ [6]
= SO_4^{2-}, SeO_4^{2-},[9] and
Y = $CH_3CO_2^-$ [6,18] or any other carboxylic acid[7,8]
= CO_3^{2-}.[10]

The reaction of the tri-μ-hydroxo-bis[triamminecobalt(III)] cation with carboxylic acids has proven to be most versatile, yielding the μ-carboxylatodi-μ-hydroxo-bis[triamminecobalt(III)] cations.[8] These binuclear complexes represent a useful class of compounds for studies of electron-transfer reactions through extended organic structural units.[10,11] Inner-sphere reductions of these complexes by aquametal ions such as $[Cr(OH_2)_6]^{2+}$ or, to a lesser extent $[V(OH_2)_6]^{2+}$, occur exclusively by way of remote attack of the reductant if the organic ligand bears a suitable "lead in" function (e.g., a carboxyl or aldehyde group), and if the ligand is capable of "mediating" an electron.[12–15] In all other cases, much slower outer-sphere reductions are observed.[16,17,19] Thus, in many instances, kinetic studies of the reductions of μ-carboxylato-di-μ-hydroxo-bis[triamminecobalt(III)] complexes complement those using mononuclear carboxylato-pentamminecobalt(III) complexes.

Furthermore, the kinetics of the OH-bridge cleavage and formation have been studied in detail.[3,20,21]

A. *mer*-TRIAMMINETRINITROCOBALT(III)

$$4\ CoCl_2{\cdot}6H_2O + 8\ NH_3 + 4\ NH_4Cl + 12\ NaNO_2 + O_2 \longrightarrow 4\ mer\text{-}[Co(NH_3)_3(NO_2)_3] + 12\ NaCl + 26\ H_2O$$

Concentrated aqueous ammonia (20–25%) (2000 mL) is added to a solution of 400 g of ammonium chloride and 540 g of sodium nitrite in 3000 mL of water. Cobalt(II) chloride hexahydrate, 360 g, dissolved in 1000 mL of water is added and a rapid current of air is sucked through this solution for a period of five hours. The dark brown solution is placed in four to five large porcelain evaporating dishes (of the approximate diameter 250 mm) and is allowed to stand in a drafty fume cupboard for two days, during which time brown crystals grow slowly on the surface of the solution. From time to time the surface is disturbed by stirring with a glass rod, which causes the crystals to sink to the bottom of the dishes. It is important for good yields that the surface of the solution be relatively large (~2000 cm^2) and that a steady stream of air passes over the surface (the temperature of the air should not exceed 20°*). The brown precipitate is collected on a filtering crucible, washed successively with 200 mL of ice water, ethanol, and diethyl ether, and air-dried. The yield of the brown precipitate (230–260 g) is found to vary from batch to batch. The crude product consists of different complexes of cobalt(III), all of which have a ratio $Co:NH_3:NO_2 = 1:3:3$ (*trans-trans*-$[Co(NH_3)_4(NO_2)_2][Co(NH_3)_2(NO_2)_4]$ and *mer*-$[Co(NH_3)_3(NO_2)_3]$ are the main products). It is the neutral complex, H *mer* $[Co(NH_3)_3(NO_2)_3]$, which is most soluble in water. The powdered, crude product is extracted with 1500 mL of water at 85°, to which 15 mL of glacial acetic acid is added. This procedure is repeated four times. The remaining yellow powder is discarded. The combined filtrates are cooled to 0°. Yellow crystals precipitate and are collected on a filtering crucible, washed successively with ice-cold water, ethanol, and diethyl ether, and then air-dried. The yields vary again (160–180 g).

This yellow product is still contaminated with *trans, trans*-$[Co(NH_3)_4(NO_2)_2][Co(NH_3)_2(NO_2)_4]$ (up to 13%). Even repeated recrystallizations do not afford a pure sample of *mer*-$[Co(NH_3)_3(NO_2)_3]$. A procedure to obtain a pure sample of *mer*-$[Co(NH_3)_3(NO_2)_3]$ is reported elsewhere.[4] For the present purpose the crude sample suffices.

*At lower temperatures, 15–17°, better yields have been obtained. It should be noted that *slow* evaporation of NH_3 appears to be the crucial factor, not the overall reduction of the volume of the solution.

B. *mer*-TRIAMMINEAQUADICHLOROCOBALT(III) CHLORIDE

$$mer\text{-}[Co(NH_3)_3(NO_2)_3] + H_2O + 3\ HCl \longrightarrow [Co(NH_3)_3(H_2O)Cl_2]Cl + 3\ HNO_2$$

Procedure

The crude H *mer*-$[Co(NH_3)_3(NO_2)_3]$ (160 g) is suspended in 1300 mL of concentrated hydrochloric acid at 20° (**Caution.** *Effervescence!*), whereupon fumes of nitrogen oxides are evolved. (*Hood!*) The yields are lower when the reaction is carried out at elevated temperatures. The solution is stirred occasionally. A gray-green precipitate is generated within 20 hours, which is then collected on a filtering crucible, washed successively with 20% hydrochloric acid, ethanol, and diethyl ether, and air-dried. The crude product is dissolved in 1500 mL of 1% aqueous hydrochloric acid and filtered. To the filtrate 1500 mL of ice-cold concentrated hydrochloric acid is added in portions of 150 mL at 0°. After 12 hours at 0°, black-violet crystals precipitate. They are filtered off, washed successively with 20% hydrochloric acid, ethanol, and diethyl ether, and air-dried. The product is contaminated with $[Co(NH_3)_4(NO_2)Cl]Cl$ (~2–3%). The yield is 129 g.

C. *fac*-TRI-μ-HYDROXO-BIS[TRIAMMINECOBALT(III)] PERCHLORATE DIHYDRATE

$$2\ H\ mer\text{-}[Co(NH_3)_3(H_2O)Cl_2]Cl + 3\ OH^- + 3\ ClO_4^- \longrightarrow [Co_2(NH_3)_6(OH)_3](ClO_4)_3\ 2H_2O + 6Cl^-$$

Procedure

■ **Caution.** *Perchlorates may explode, especially when heated!*

To an ice-cold suspension of 100 g of the crude H *mer*-$[Co(NH_3)_3(H_2O)Cl_2]Cl$ in 100 mL of water, 600 mL of 1 *M* aqueous ammonia is added dropwise over a period of two hours with efficient stirring. The temperature must be kept at 0°. A turbid, deep red solution is obtained which is filtered.* To this solution 250 mL of a saturated aqueous solution of sodium perchlorate is added. It is then cooled to 0° for one hour. The red precipitate is collected on a filtering crucible and recrystallized from 800 mL of water (weakly acidified with acetic acid) at 30° by adding 200 mL of a saturated aqueous solution of $NaClO_4$ and cooling to 0° for three hours. Red, hexagonal prismatic crystals are filtered off,

*The red residue (~5 g) consists mainly of $[Co(NH_3)_4(NO_2)Cl]Cl$ and a little $[Co(NH_3)_5Cl]Cl_2$.

washed with ethanol, and then with diethyl ether, and air-dried. The yield is 81 g (63%). The product is slightly contaminated with chloride, which does not affect the following syntheses. *Anal.* Calcd. for $[(NH_3)_3Co(OH)_3Co(NH_3)_3](ClO_4)_3 \cdot 2H_2O$: Co, 19.47; NH_3, 16.88; H_2O, 5.95; ClO_4, 49.28. Found: Co, 19.4; NH_3, 16.9; H_2O, 6.1; ClO_4, 47.7.

Properties

The product can be characterized by two absorption maxima in the visible region. λ_{max} = 526 nm (ϵ = 134 l mol^{-1}cm^{-1}), 296 nm (ϵ = 1.78 × 10^3 l mole^{-1}cm^{-1}). The dithionate salt has been characterized by single crystal X-ray analysis, which shows the binuclear cation to consist of two face sharing octahedra.[2] The cation is only moderately stable in aqueous solution (pH 7), but at pH 5–6 aqueous solutions can be kept at ambient temperature for many hours without decomposition.[3]

D. DI-μ-HYDROXO-DIAQUABIS[TRIAMMINECOBALT(III)] TETRAPERCHLORATE PENTAHYDRATE

$$[Co_2(NH_3)_6(OH)_3](ClO_4)_3 \cdot 2H_2O + 4H_2O + HClO_4 \longrightarrow [Co_2(NH_3)_6(OH)_2(H_2O)_2](ClO_4)_4 \cdot 5H_2O$$

Tri-μ-hydroxo-bis[triamminecobalt(III)] triperchlorate dihydrate, 10 g (16.5 mmole) is dissolved with stirring in 100 mL of 1 *M* perchloric acid at 30°. After 15 minutes, 200 mL of 60% perchloric acid is added in small amounts to the filtered solution. The solution is kept at 0° for four hours, after which time red-violet crystals are collected on a filtering crucible. The product is washed with a 1 : 1 mixture of ethanol and diethyl ether, and then pure diethyl ether, and air-dried. The yield is 9.0 g (70%). *Anal.* Calcd. for $[Co_2(NH_3)_6(H_2O)_2(OH)_2](ClO_4)_4 \cdot 5H_2O$: Co, 15.15; NH_3, 13.13; ClO_4, 51.13. Found: Co, 15.0; NH_3, 13.2; ClO_4, 50.9.

Properties

The perchlorate salt is very soluble in water and moderately so in 96% ethanol. The UV-absorption spectrum has maxima at λ = 544 nm (ϵ = 170 l mole^{-1}cm^{-1}), 295 nm (ϵ = 2.69 × 10^3 l mole^{-1}cm^{-1}). A single-crystal X-ray analysis of the nitrate salt, $[(NH_3)_6Co_2(OH)_2(OH_2)_2]$ $(NO_3)_4 \cdot 2\ H_2O$, shows the cation to have two edge-sharing octahedra. The two coordinated aqua ligands are in *trans* positions with respect to each other.[22]

Prolonged reaction of tri-μ-hydroxo-bis[triamminecobalt(III)] perchlorate dihydrate with concentrated perchloric acid yields the monomeric complex, $[Co(NH_3)_3(H_2O)_3](ClO_4)_3$.[3]

E. μ-CARBOXYLATODI-μ-HYDROXO-BIS[TRIAMMINECOBALT(III)] PERCHLORATE

$$[Co_2(NH_3)_3(OH)_3](ClO_4)_3 \cdot 2\ H_2O + R\text{—}COOH \longrightarrow$$
$$[Co_2(NH_3)_6(OH)_2R\text{—}COO](ClO_4)_3 + 3\ H_2O$$

Procedure

■ **Caution.** *Perchlorates may explode, especially when heated! Perform this experiment behind a safety shield.*

A general procedure is given for the reaction of various water-soluble carboxylic acids with tri-μ-hydroxo-bis[triamminecobalt(III)] perchlorate. To a solution of 20 mmoles of the carboxylic acid in 40 mL of 0.5 *M* perchloric acid at 45°, 6.1 g (10 mmoles) of tri-μ-hydroxo-bis[triamminecobalt(III)] perchlorate dihydrate is added in small amounts with stirring. The temperature is raised to 65° **(Caution)** and maintained for 15 minutes. To the cooled (15°) and filtered solution, 10 mL of concentrated perchloric acid (70%) and 2 g of solid sodium perchlorate are added. The solution is kept at 0° for three hours. Red crystals are filtered off and recrystallized from 20 mL of water at 40° by adding 20 mL of concentrated perchloric acid. The temperature is maintained at 0° for three hours. The yields vary: 30–90%. The following μ-carboxylato complexes can be prepared according to the above procedure:

1. μ-Acetatodi-μ-hydroxo-bis[triamminecobalt(III)] Triperchlorate Dihydrate*

```
[                OH                         ]
[              /    \                       ]
[ (NH3)3Co —— OH —— Co(NH3)3                ]
[          \            /                   ]
[           O . . . . O                     ]
[              \    /                       ]
[                C                          ]
[                |                          ]
[               CH3                         ]
```

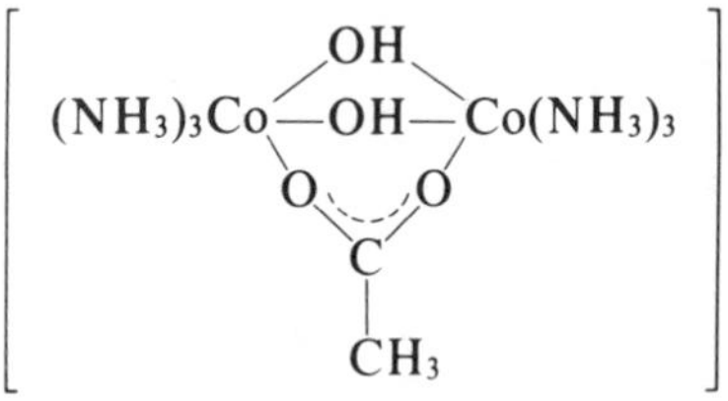

$(ClO_4)_3 \cdot 2H_2O$. The yield is 2.8 g (44%).

*In the original paper[6] a double salt is obtained, $[(NH_3)_6Co_2(OH)_2(CH_3CO_2)]\ (ClO_4)_3 \cdot 0.9\ NaClO_4$ using $NaClO_4$ to precipitate red crystals.

Anal. Calcd. for $[(NH_3)_6Co_2(OH)_2(CH_3CO_2)]\ (ClO_4)_3 \cdot 2\ H_2O$: Co, 18.20; N, 12.97; C, 3.71; H, 4.20; Cl, 16.42. Found: Co, 18.1; N, 12.8; C, 3.9; H, 4.3; Cl, 16.3.

2. μ-Hydrogenoxalatodi-μ-hydroxo-bis[triamminecobalt(III)] Triperchlorate Hemihydrate[7]

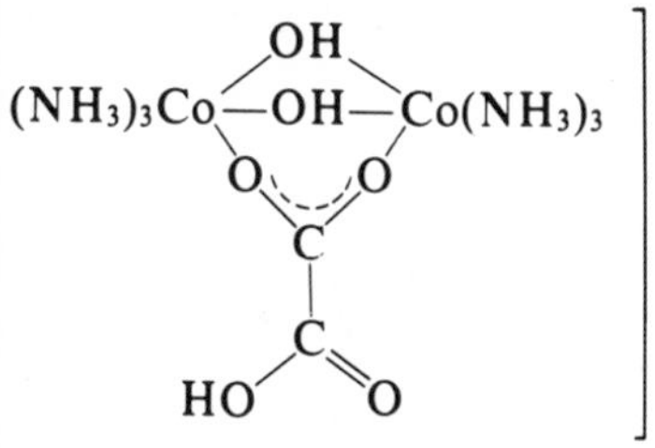

OH
$(NH_3)_3Co$—OH—$Co(NH_3)_3$
O O
C
C
HO O

$(ClO_4)_3 \cdot 1/2\ H_2O$. The yield is 4.1 g (63%).

Anal. Calcd. for $[(NH_3)_6Co_2(OH)_2(C_2O_4H)](ClO_4)_3 \cdot 0.5\ H_2O$: Co, 18.12; NH_3, 15.71; Cl, 16.35; C, 3.69; H, 3.41. Found: Co, 18.2; NH_3, 15.6; Cl, 16.1; C, 3.9; H, 3.6.

3. μ-(4-Pyridinecarboxylato)di-μ-hydroxo-bis[triamminecobalt(III)] Tetraperchlorate[8]

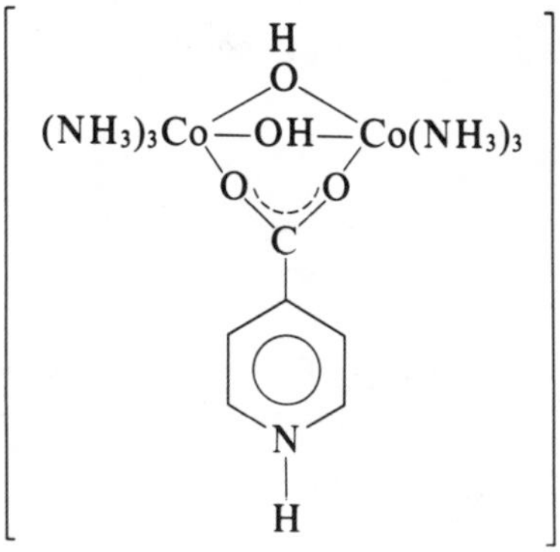

H
O
$(NH_3)_3Co$—OH—$Co(NH_3)_3$
O O
C
N
H

$(ClO_4)_4$. The yield is 7.1 g (92%).

Anal. Calcd. for $[(NH_3)_6Co_2(OH)_2(C_5H_4NHCOO)](ClO_4)_4$: Co, 15.2; NH_3, 13.2; C, 9.3; H, 3.3; Cl, 18.3. Found: Co, 15.1; NH_3, 13.1; C, 9.5; H, 3.4; Cl, 18.4.

4. μ-Pyrazinecarboxylato-di-μ-hydroxo-bis[triamminecobalt(III)] Tetraperchlorate Hydrate[8]

Conc. $HClO_4$ (30 mL) was added to precipitate red crystals within 12 hours at 0°. The yield is 3.5 g (45%).

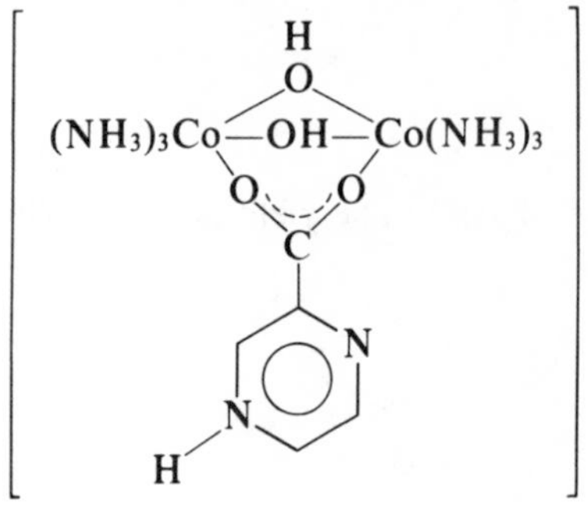

$[(NH_3)_3Co(OH)_2(C_4H_4N_2COO)Co(NH_3)_3](ClO_4)_4 \cdot H_2O$. The yield is 3.5 g (45%).

Anal. Calcd. for $[(NH_3)_6Co_2(OH)_2(C_4H_4N_2COO)](ClO_4)_4 \cdot H_2O$: Co, 14.8; NH_3, 12.9; C, 7.6; H, 3.3; Cl, 17.9. Found: Co, 14.7; NH_3, 12.6; C, 7.8; H, 3.2; Cl, 17.8.

A slight modification of the above procedure, allowing 5 mmoles of a water soluble dicarboxylic acid and 12 mmoles of tri-μ-hydroxo-bis-[triamminecobalt(III)]perchlorate dihydrate to react, affords the following tetranuclear complexes:

5. Tetra-μ-hydroxo-μ$_4$-oxalato-tetrakis[triamminecobalt(III)] Hexaperchlorate Tetrahydrate[6]

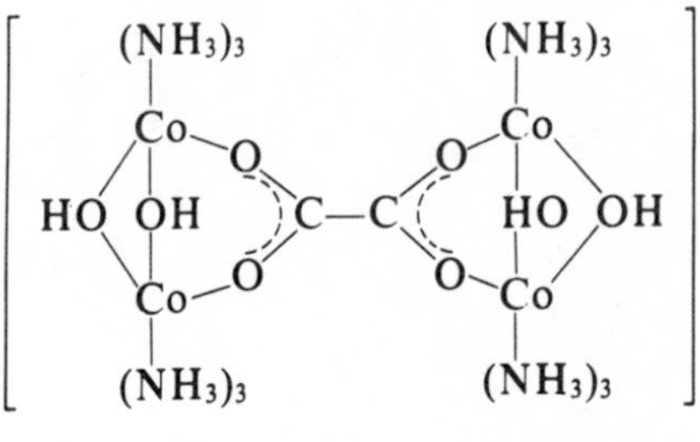

$[\{(NH_3)_3Co(OH)_2Co(NH_3)_3\}_2(C_2O_4)](ClO_4)_6 \cdot 4H_2O$. The yield is 4.1 g (65% (based on oxalic acid).

Anal. Calcd. for $[(NH_3)_{12}Co_4(OH)_4(C_2O_4)]$ $(ClO_4)_6 \cdot 4\ H_2O$: Co, 18.64; NH_3, 16.16; C, 1.90; H, 3.83; Cl, 16.82. Found: Co, 18.6; NH_3, 16.0; C, 2.2; H, 3.9; Cl, 16.8.

6. μ_4-Acetylenedicarboxylato-tetra-μ-hydroxo-tetrakis[triamminecobalt(III)] Hexaperchlorate Pentahydrate[12]

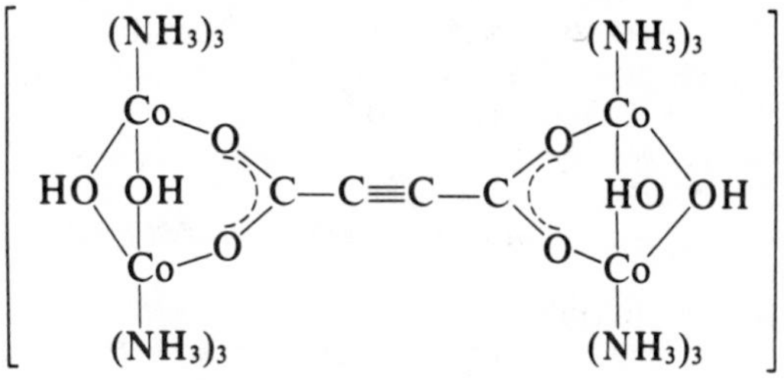

$$\left[(NH_3)_3Co(OH)_2(O_2C{-}C{\equiv}C{-}CO_2)Co(NH_3)_3 \cdots \right]$$

$(ClO_4)_6 \cdot 5H_2O$.
The yield is 3.5 g (54%) (based on acetylenedicarboxylic acid).

Anal. Calcd. for $[(NH_3)_{12}Co_4(OH)_4(C_4O_4)]\ (ClO_4)_6{\cdot}5\ H_2O$: Co, 18.0; N, 12.9; C, 3.7; H, 3.8; Cl, 16.3. Found: Co, 17.7; N, 12.6; C, 3.8; H, 4.0; Cl, 16.2.

Properties

The binuclear and tetranuclear μ-carboxylato complexes are remarkably substitution-inert in acidic aqueous solutions (pH 5–0). The dinuclear structure has been established by single-crystal X-ray analyses in two instances.[18,19] In the case of the μ-pyrimidine-carboxylato complex, a symmetrical μ-carboxylato bridged group is found.[19]

The μ-carboxylato complexes described here have very similar absorption spectra in the visible region. The $[(NH_3)_6Co_2(OH)_2(R\text{-}COO)]$ structural unit exhibits an absorption maximum at $\lambda_{max} = 524$ nm ($\epsilon = 109 \pm 3$ l mole^{-1}cm^{-1}) regardless of different basicities of the various carboxylate ligands. A linear correlation between the γ_{as}(C—O) stretching frequency of the bridging carboxylate and pK_a values for the dissociation constants of the free acids is observed.[8]

References

1. (a) A. Werner, *Ber.*, **40,** 4834 (1907); (b) *Ibid.*, **41,** 3879 (1908).
2. U. Thewalt, *Z. anorg. allg. Chem.*, **412,** 29 (1975).
3. M. Linhard and H. Siebert, *Z. anorg. allg. Chem.*, **364,** 24 (1969).
4. H. Siebert, *Z. anorg. allg. Chem.*, **441,** 47 (1978).
5. S. M. Jörgensen, *Z. anorg. Chem.*, **17,** 475 (1898).
6. H. Siebert and G. Tremmel, *Z. anorg. allg. Chem.*, **390,** 292 (1972).
7. K. Wieghardt, *Z. anorg. allg. Chem.*, **391,** 142 (1972).
8. K. Wieghardt, *J. Chem. Soc. Dalton Trans.*, 2548 (1973).
9. (a) K. Wieghardt and J. Eckert, *Z. anorg. allg. Chem.*, **383,** 241 (1971); (b) K. Wieghardt and G. Maas, *Z. anorg. allg. Chem.*, **385,** 289 (1971).
10. M. R. Churchill, R. A. Lashewycz, K. Koshy, and T. P. Dasgupta, *Inorg. Chem.*, **20,** 376 (1981).

11. H. Taube and E. S. Gould, *Accounts Chem. Res.*, **2,** 321 (1969).
12. M. Hery and K. Wieghardt, *J. Chem. Soc. Dalton Trans.*, 1536 (1976).
13. H. Spiecker and K. Wieghardt, *Inorg. Chem.*, **16,** 1290 (1977).
14. H. Spiecker and K. Wieghardt, *Inorg. Chem.*, **15,** 909 (1976).
15. I. Baldea, K. Wieghardt, and A. G. Sykes, *J. Chem. Soc. Dalton Trans.*, 78 (1977). K. L. Scott, K. Wieghardt, and A. G. Sykes, *Inorg. Chem.*, **12,** 655 (1973).
16. K. L. Scott and A. G. Sykes, *J. Chem. Soc. Dalton Trans.*, 1832 (1972).
17. (a) K. Wieghardt and A. G. Sykes, *J. Chem. Soc. Dalton Trans.*, 651 (1974); (b) H. M. Huck and K. Wieghardt, *Inorg. Chem.*, **19,** 3688 (1980).
18. G. S. Mandel, R. E. Marsh, W. P. Schaefer, and N. S. Mandel, *Acta Cryst.*, **B33,** 3185 (1977).
19. G. Maas, *Z. anorg. allg. Chem.*, **432,** 203 (1977).
20. J. D. Edwards, K. Wieghardt, and A. G. Sykes, *J. Chem. Soc. Dalton Trans.*, 2198 (1974).
21. W. Jentsch, W. Schmidt, A. G. Sykes, and K. Wieghardt, *Inorg. Chem.*, **16,** 1935 (1977).
22. W. H. Baur and K. Wieghardt, *J. Chem. Soc. Dalton Trans.*, 2669 (1973).

22. TETRABUTYLAMMONIUM OCTACHLORODIRHENATE(III)

$$2[(n\text{-}C_4H_9)_4N]ReO_4 + 8C_6H_5COCl \longrightarrow [(n\text{-}C_4H_9)_4N]_2[Re_2Cl_8] + \text{Org. Prod.}$$

Submitted by T. J. BARDER* and R. A. WALTON*
Checked by F. A. COTTON† and G. L. POWELL†

The octachlorodirhenate(III) anion is the key starting material for entry into the chemistry of multiply bonded complexes of dirhenium.[1] To date, the most convenient synthetic route for the synthesis of this complex has been through the hypophosphorous acid reduction of $KReO_4$.[2,3] However, the yield of the complex is low and variable, rarely exceeding 40%. Other synthetic routes[3–6] require either high pressure conditions or the use of starting materials that are best prepared from the $[Re_2Cl_8]^{2-}$ anion itself, and they offer no significant advantages over the hypophosphorous acid reduction of $KReO_4$. We describe here a simple, quick, and high-yield synthesis of $[(n\text{-}C_4H_9)_4N]_2[Re_2Cl_8]$, which has many advantages over existing procedures.[7]

*Department of Chemistry, Purdue University, West Lafayette, IN 47907.
†Department of Chemistry, Texas A & M University, College Station, TX 77843.

Procedure

The starting material $[n\text{-}C_4H_9)_4N]ReO_4$* (3.0 g, 6.1 mmoles) is placed in a 250-mL round-bottomed flask fitted with a reflux condenser. The condenser is connected to a mercury bubbler system, which consists of a cylindrical glass reservoir (22 mm internal diameter) containing 40 mL of mercury and fitted with a 4-mm diameter gas inlet tube extending 95 mm into the mercury pool. The reaction vessel is purged with N_2 and benzoyl chloride† (30 mL, 26 mmoles) is then syringed into the reaction flask. The resulting mixture is refluxed for 1.5 hours under a positive pressure of N_2. Under these conditions, the boiling point of the benzoyl chloride should be very close to 209°.‡ The resulting dark green solution is allowed to cool and is then treated with a solution of $[(n\text{-}C_4H_9)_4N]Br$ (5.0 g, 16 mmoles) in ethanol (75 mL) which has been saturated with hydrogen chloride gas. This mixture is then refluxed for an additional hour, and the resulting solution is evaporated to approximately one half its original volume under a stream of N_2. The resulting blue crystals are filtered off, washed with three × 10 mL portions of ethanol, then with diethyl ether, and dried under vacuum. Yield: 94% (3.28 g). *Anal.* Calcd. for $C_{32}H_{72}Cl_8N_2Re_2$: C, 33.69; H, 6.36. Found: C, 33.97; H, 6.12.

Properties

The complex $[(n\text{-}C_4H_9)_4N]_2[Re_2Cl_8]$ is soluble in acetone, acetonitrile, methanol, and many other nonaqueous solvents. It can be recrystallized by dissolution of the complex in boiling methanol, followed by filtration into an equal volume of concentrated HCl. Subsequent evaporation of the solution to remove the methanol gives dark blue crystals. The complex is quite stable in air and can be stored indefinitely without special precautions.

This complex possesses a metal–metal quadruple bond, and can be converted readily into other dirhenium species containing multiple bonds. Solutions of $[Re_2Cl_8]^{2-}$ react with concentrated HBr to form the analogous bromo-anion $[Re_2Br_8]^{2-}$,[2] and with acetic acid/acetic anhydride mixtures to produce the acetate-bridged dirhenium(III) complex $Re_2(O_2CCH_3)_4Cl_2$.[8] The $[(n\text{-}C_4H_9)_4N]_2[Re_2Cl_8]$ complex reacts with phosphines (PR_3) to yield complexes of stoichiometry $Re_2Cl_6(PR_3)_2$, $Re_2Cl_5(PR_3)_3$, and $Re_2Cl_4(PR_3)_4$.[9]

*This salt can be prepared in essentially quantitative yield by the slow addition of a hot, aqueous solution of $[(n\text{-}C_4H_9)_4N]Br$ to one of $KReO_4$. The resulting white precipitate is washed with water and dried in vacuum.

†Benzoyl chloride can be used as received from a commercial source.

‡This is a critical factor in ensuring a high yield of the desired product. For example, with a mineral oil bubbler, the bp of benzoyl chloride was ~198°, and the yield of $[(n\text{-}C_4H_9)_4N]_2Re_2Cl_8$ fell from ~90% to 60%.

References

1. F. A. Cotton and R. A. Walton, *Multiple Bonds Between Metal Atoms*, Wiley-Interscience, New York, 1982.
2. F. A. Cotton, N. F. Curtis, and W. R. Robinson, *Inorg. Chem.*, **4,** 1696 (1965).
3. F. A. Cotton, N. F. Curtis, B. F. G. Johnson, and W. R. Robinson, *Inorg. Chem.*, **4,** 326 (1965).
4. A. S. Kotel'nikova, M. I. Glinkina, T. V. Misatlova, and V. G. Lebedev, *Russ. J. Inorg. Chem.*, **21,** 547 (1976).
5. R. A. Bailey and J. A. McIntyre, *Inorg. Chem.*, **5,** 1940 (1966).
6. A. B. Brignole and F. A. Cotton, *Inorg. Synth.*, **13,** 82 (1972).
7. T. J. Barder and R. A. Walton, *Inorg. Chem.*, **21,** 2510 (1982).
8. F. A. Cotton, C. Oldham, and W. R. Robinson, *Inorg. Chem.*, **5,** 1798 (1966).
9. J. R. Ebner and R. A. Walton, *Inorg. Chem.*, **14,** 1987 (1975).

23. MOLYBDENUM-SULFUR CLUSTER CHELATES

Submitted by H. KECK,* W. KUCHEN,* and J. MATHOW*
Checked by W. K. MILLER,† L. TANNER,† L. L. WRIGHT,† C. J. CASEWIT,† and M. RAKOWSKI DUBOIS†

The coordination chemistry of molybdenum is of great importance because of its involvement in a variety of biological processes.[1] For example, molybdenum is responsible for the activity of some enzymes (e.g., nitrogenase). Such enzymes catalyze redox reactions like the enzymatic dinitrogen reduction. It is known that sulfur-containing ligands play an important role in such processes, and all known molybdenum-containing enzymes have S-ligand functions.[1] Recently, the syntheses of the Mo-S-cluster-chelates $[Mo_3(Et_2PS_2)_3(S_2)_3S][Et_2PS_2]$ and $[Mo_3(Et_2PS_2)_4S_4]$, containing the cluster units $Mo_3S_7^{2}$ (Fig. 1) and $Mo_3S_4^{2}$ (Fig. 2), respectively, and dithiophosphinato-chelate ligands were reported, and some of their spectroscopic and chemical properties were described.[3,4] The methods given here provide a convenient access to $[Mo_3S_7]$ and $[Mo_3S_4]$ clusters modified by phosphorus and organo groups. Thus, not only good solubility in organic solvents is achieved, but functionalization of these compounds by suitable choice of organo substituents seems also to be possible. The structures of both clusters have been determined by X-ray analysis. The $[Mo_3(Et_2PS_2)_4S_4]$ complex, containing sulfur in four different bonding states, is the first trinuclear Mo-S-cluster

*Institut für Anorganische Chemie und Strukturchemie I, University of Düsseldorf, 4000 Düsseldorf, Federal Republic of Germany.
†Department of Chemistry, University of Colorado, Boulder, CO 80309.

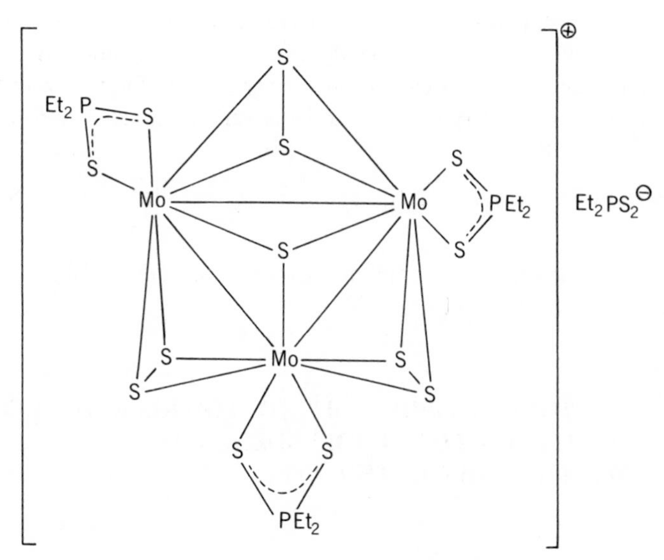

Fig. 1. Cluster unit Mo_3S_7

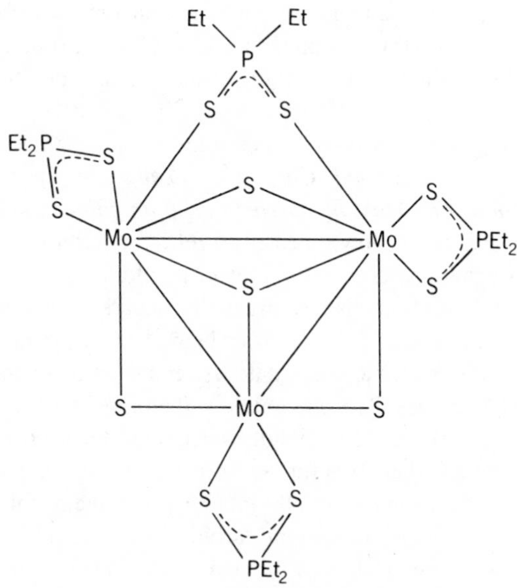

Fig. 2. Cluster unit Mo_3S_4

containing a coordinatively unsaturated metal atom to be reported. It readily adds pyridine with maintenance of the Mo_3 triangular configuration. Both clusters can easily be converted into each other by partial desulfuration with Ph_3P or reaction with S_8 or $Et_2P(S)S_2P(S)Et_2$, the latter serving as sulfur transfer agent.

$$[Mo_3(Et_2PS_2)_3(S_2)_3S][Et_2PS_2] \underset{+S_8 \text{ or } Et_2P(S)S_2P(S)Et_2}{\overset{+Ph_3P-Ph_3PS}{\rightleftarrows}} [Mo_3(Et_2PS_2)_4S_4] \quad (1)$$

The compounds described are useful materials for the study of the chemical and spectroscopic behavior of trinuclear Mo-S-clusters.

A. TRIS(DIETHYLPHOSPHINODITHIOATO)TRIS-μ-DISULFIDO-μ_3-THIO-*TRIANGULO*-TRIMOLYBDENUM(IV) DIETHYLPHOSPHINODITHIOATE

$$3\ Mo(CO)_6 + 9\ Et_2P(S)S_2P(S)Et_2 \longrightarrow [Mo_3(Et_2PS_2)_3(S_2)_3S][Et_2PS_2] + 7\ Et_2P(S)SP(S)Et_2 + 18\ CO$$

Procedure

A 250-mL two-necked, round-bottomed flask fitted with an argon-inlet tube and a reflux condenser with a silicone oil bubbler is charged with $Mo(CO)_6$ (5.3 g, 0.02 mole), dithiobis(diethylphosphine sulfide)[5] (18.4 g, 0.06 mole), 70 mL of toluene (dried by distillation from P_4O_{10}), and a magnetic stirring bar coated with Teflon.

■ **Caution.** *Metal carbonyl compounds are toxic chemicals and should be handled with care in a hood. Carbon monoxide is evolved in the reaction, so that preparation, too, must be carried out in an efficient hood.*

The reaction mixture is slowly heated to reflux under a stream of argon and with magnetic stirring. After a few minutes, a colorless, clear solution forms, and the color of the solution changes gradually to dark brown, and evolution of carbon monoxide starts at about 80°. After being heated for an additional hour, a yellow-orange solid begins to precipitate. After a reaction period of four hours, the evolution of CO ceases. The precipitate is then filtered off in air and washed three times with 20-mL portions of hot toluene and then three times with 20-mL portions of diethyl ether. It is further purified by dissolving it in a minimum volume of dichloromethane (about 300 mL) to which methanol is slowly added until the solution just begins to become turbid (about 200 mL). After filtration, the clear yellow solution is allowed to stand in an open Erlenmeyer flask for some hours, producing yellow-orange crystals, which are dried in vacuum (6.2

g, yield 85%). *Anal.* Calcd. for $C_{16}H_{40}P_4S_{15}Mo_3$: C, 17.08; H, 3.58; P, 11.01; S, 42.74; Mo, 25.60. Found: C, 17.04; H, 3.51; P, 11.04; S, 42.32; Mo, 25.91.

Properties

The compound forms yellow-orange, diamagnetic crystals, which begin to decompose at about 260°. They are stable in moist air and slightly soluble in chloroform and dichloromethane. The $^{31}P(^{1}H)$ nmr spectrum (saturated solution in CH_2Cl_2, 85% H_3PO_4 as reference) shows singlets at δ_P = 110.1 ppm (chelate ligands) and 73.5 ppm (anion). The FD mass spectrum shows signals for the cluster cation $[Mo_3(Et_2PS_2)_3S_7]^+$ (m/z 977, ref. to ^{98}Mo, ^{32}S). The compound is a 1:1 electrolyte with a specific conductivity $\Lambda_c = 2.5 \times 10^{-3}\ \Omega^{-1}\ mole^{-1}cm^2$ (0.0015 *M*, CH_2Cl_2, 25°). A single crystal X-ray structure determination has been published.[3]

B. μ-(DIETHYLPHOSPHINODITHIOATO)-TRIS(DIETHYLPHOSPHINODITHIOATO)-TRI-μ-THIO-μ$_3$-THIO-*TRIANGULO*-TRIMOLYBDENUM(IV)

$$[Mo_3(Et_2PS_2)_3(S_2)_3S][Et_2PS_2] + 3\ Ph_3P \longrightarrow [Mo_3(Et_2PS_2)_4S_4] + 3\ Ph_3PS$$

Procedure

A 250-mL round-bottomed flask fitted with a reflux condenser is charged with $[Mo_3(Et_2PS_2)_3(S_2)S][Et_2PS_2]$ (5.0 g, 0.0044 mole), Ph_3P (3.5 g, 0.0132 mole), 100 mL of dichloromethane (freshly distilled) and a magnetic stirring bar coated with Teflon. The reaction mixture is stirred at room temperature for one hour. Then the solvent is removed at room temperature in a rotatory evaporator under reduced pressure. The black residue is washed four times with 20-mL portions of hot methanol and then recrystallized from 100 mL toluene, affording bright black crystals, which are dried in vacuum (4.5 g, yield 98%). *Anal.* Calcd. for $C_{16}H_{40}P_4S_{12}Mo_3$: C, 18.67; H, 3.92; P, 12.04; S, 37.39; Mo, 27.97. Found: C, 18.65; H, 3.89; P, 12.00; S, 37.00; Mo, 27.69.

Properties

The compound is stable toward air and moisture and is soluble in benzene, toluene, chloroform, and dichloromethane. On heating, the bright, black, diamagnetic crystals begin to decompose at about 195°. The structure of the complex has been established by a single-crystal X-ray determination.[4] Its FD mass spectrum shows $[M]^+$ at m/z 1034 (ref. to $^{98}Mo^{32}S$). The $^{31}P(^{1}H)$ spectrum of the substance (0.03 *M* in CH_2Cl_2) shows singlets at δ_P = 123.2, 114.8, and 83.7 ppm (intensity 1:2:1, 85% H_3PO_4 as reference). One of the three molybdenum atoms in the cluster is coordinatively unsaturated and can add one molecule of pyridine.[4] It should be noticed that, on treating the compound with S_8 or $Et_2P(S)S_2P(S)Et_2$ (as a sulfur donor), the yellow-orange starting compound $[Mo_3(Et_2PS_2)_3(S_2)_3S][Et_2PS_2]$ can be regenerated.[4]

References

1. E. I. Stiefel, *Prog. Inorg. Chem.*, **22,** 1 (1977).
2. (a) A. Müller, R. Jostes, and F. A. Cotton, *Angew. Chem.*, **92,** 921 (1980); (b) *Angew. Chem. Int. (Engl. Ed.)*, **19,** 875 (1980).
3. (a) H. Keck, W. Kuchen, J. Mathow, B. Meyer, D. Mootz, and H. Wunderlich, *Angew. Chem.*, **93,** 1019 (1981); (b) *Angew. Chem. Int. (Engl. Ed.)*, **20,** 975 (1981).
4. (a) H. Keck, W. Kuchen, J. Mathow, and H. Wunderlich, *Angew. Chem.*, **94,** 927 (1982); (b) *Angew. Chem. Int. (Engl. Ed.)*, **21,** 929 (1982).
5. W. Kuchen, K. Strolenberg, and J. Metten, *Chem. Ber.*, **96,** 1733 (1963).

24. DIRHODIUM(I) COMPLEXES WITH CARBON MONOXIDE AND SUBSTITUTED NITROGEN, PHOSPHORUS, AND SULFUR LIGANDS

Submitted by PHILIPPE KALCK,* PIERRE-MARIE PFISTER,* TIMOTHY G. SOUTHERN,* and ALAIN THOREZ*
Checked by L. CHEN,† D. W. MEEK,† C. WOODCOCK,‡ and R. EISENBERG‡

$$[Rh_2(\mu\text{-}Cl)_2(\eta^4 - C_8H_{12})_2] + 4\ P(OCH_3)_3 \longrightarrow [Rh_2(\mu\text{-}Cl)_2\{P(OCH_3)_3\}_4] + 2\ C_8H_{12}$$

$$[Rh_2(\mu\text{-}Cl)_2\{P(OCH_3)_3\}_4] + 2\ Li(t\text{-}C_4H_9S) \longrightarrow [Rh_2(\mu\text{-}t\text{-}C_4H_9S)_2\{P(OCH_3)_3\}_4] + 2\ LiCl.$$

$$[Rh_2(\mu\text{-}Cl)_2(CO)_4] + 2\ t\text{-}C_4H_9SH \longrightarrow [Rh_2(\mu\text{-}t\text{-}C_4H_9S)_2(CO)_4] + 2\ HCl$$

$$[Rh_2(\mu\text{-}t\text{-}C_4H_9S)_2(CO)_4] + 2\ P(OCH_3)_3 \longrightarrow [Rh_2(\mu\text{-}t\text{-}C_4H_9S)_2(CO)_2\{P(OCH_3)_3\}_2] + 2\ CO$$

*Laboratoire de chimie minérale et de cristallochimie, Ecole Nationale Supérieure de Chimie, Toulouse, France 31077.
†Department of Chemistry, The Ohio State University, Columbus, OH 43210.
‡Department of Chemistry, The University of Rochester, Rochester, NY 14627.

The addition of 1 or 2 moles of phosphite ligand to the complex $[Rh_2(\mu\text{-}t\text{-}C_4H_9S)(CO)_2\{P(OCH_3)_3\}_2]$ does not displace the CO groups. To prepare the first complex, the special ability of the phosphite ligand, which does not cleave the chloro bridges in $[Rh_2(\mu\text{-}Cl)_2(\eta^4 - C_8H_{12})_2]$,[1] but instead displaces the cyclo-octadiene ligand, is used.

A. BIS-μ-(2-METHYL-2-PROPANETHIOLATO)-TETRAKIS(TRIMETHYL PHOSPHITE)DIRHODIUM(I)

Procedure[2]

■ **Caution.** *All the manipulations must be carried out under an efficient fume hood due to the toxicity of the volatile ligands, $t\text{-}C_4H_9$ SH, $P(OCH_3)_3$ and $1,5\text{-}C_8H_{12}$. Moreover butyllithium is flammable upon contact with air.*

A solution of 2.00 g of $[Rh_2(\mu\text{-}Cl)_2(\eta^4\text{-}C_8H_{12})_2]$* (4.06 mmoles) dissolved in 30 mL of CH_2Cl_2 at room temperature is placed into a Schlenck tube of 50 mL capacity. Then 1.82 mL of freshly distilled, dry trimethyl phosphite (16.23 mmole) is slowly added at room temperature with vigorous stirring. If a precipitate appears, the rate of addition must be reduced. Then the precipitate of $[Rh\{P(OCH_3)_3\}_4]Cl$[4,5] redissolves, giving a mixture of dinuclear complexes.[5] After addition of the ligand is complete, the solution is stirred for two hours. The solvent is slowly evapored under reduced pressure to remove the cyclooctadiene. The $[Rh_2(\mu\text{-}Cl)_2\{P(OCH_3)_3\}_4]$ complex should be obtained as a yellow powder. If an oil is observed, it is advisable either to recrystallize the complex from toluene/hexane solution, or to dissolve it in toluene or dichloromethane and evaporate the solution in order to remove the remaining cyclooctadiene.

The yellow powder is dissolved in 20 mL of toluene and 8.9 mmole of $(t\text{-}C_4H_9S)Li$ (obtained by mixing 3.99 mL of 2.23 M butyllithium† and 0.784 mL of 2-methyl-2-propanethiol) in 20 mL of toluene is added. The solution, which immediately turns to a luminescent brown-yellow color, is stirred for three hours. Instead of filtration, as has been described for the complex $[Ir_2(\mu\text{-}t\text{-}C_4H_9S)_2(\eta^4\text{-}C_8H_{12})]$,[6] centrifugation of the toluene solution allows the lithium chloride, which may contain a colloidal fraction, to be eliminated rapidly. Evaporation of toluene under reduced pressure gives a brown, viscous oil which is dissolved in the minimum volume of petroleum ether,‡ about 20 mL. Crystallization at −25°,

*Available from Alfa Products, Ventron Corp., P.O. Box 299, Danvers, MA 01923.
†Available from Aldrich Chemical Co., 940 W. St. Paul Ave., Milwaukee, WI 53233.
‡The checkers report that it is preferable to dissolve the complex in toluene and then add hexane to precipitate the complex.

after concentration of the mother liquor, gives 3.11 g of the first compound as light brown crystals. (Yield = 87%) *Anal.* Calcd. for $C_{20}H_{54}S_2O_{12}P_4Rh_2$: C, 27.28; H, 6.18; S, 7.28; P, 14.07. Found C, 27.31; H, 6.21; S, 7.23; P, 14.27.

Preparation of Related Compounds

All the complexes of general formula $[Rh_2(\mu\text{-}SR)_2\{P(OR')_3\}_4]$, where R and R′ represent an alkyl or aryl group, can be prepared by this method. However, very frequently oils or waxes are obtained, as is the case for the complexes we have checked (e.g., R = Et, Ph, CH_2Ph, *t*-C_4H_9, R′ = Me, Ph, $C_{12}H_{25}$).

Properties

Di-μ-(2-methyl-2-propanethiolato)tetrakis(trimethyl phosphite)dirhodium(I) is a light-brown, crystalline product, which is air-sensitive, even in the solid state. It gives a dark brown product upon contact with air, which is soluble in dichloromethane. The compound is soluble in many organic solvents, some alkanes included. This complex and its analogs are good catalysts for the hydrogenation of alkenes[7,8] and the hydroformylation of alkenes[8] at low temperature (80°) and pressure (2–5 bars).

B. DICARBONYL-BIS-μ-(2-METHYL-2-PROPANETHIOLATO)-BIS(TRIMETHYL PHOSPHITE)DIRHODIUM(I)

The preparations of the various complexes of the type $[Rh_2(\mu\text{-}SR)_2(CO)_2L_2]$ have been reported to result from the substitution of the chloro-bridges in $[Rh_2(\mu\text{-}Cl)_2(CO)_2L_2]$ by Li-SR.[9] When L is a phosphite ligand and R = *t*-C_4H_9, advantage is gained by the direct addition of two moles of ligand to the tetracarbonyl thiolato-bridged starting material in petroleum ether solution. The CO substitution is quantitative in a few minutes at room temperature, and crystallization of the product leads directly to high yields of pure compound.

Procedure[2,3]

■ **Caution.** *See Section A.*

A sample of 0.900 g of $[Rh_2(\mu\text{-}Cl)_2(CO)_4]$[10] * (2.31 mmoles) is dissolved in 25 mL of hexane in a 50-mL Schlenck tube with stirring. The solution is gently warmed until complete dissolution. A slight excess of 2-methyl-2-propanethiol*

*Available from Aldrich Chemical Co., Milwaukee, WI.

(0.550 mL; 4.85 mmoles) is introduced through the rubber septum, using a syringe. The mixture is warmed to about 60° for 30 minutes to achieve complete substitution of the chloro bridges.[11] The solvent, the excess of thiol, and the hydrochloric acid formed during the substitution are evaporated under reduced pressure (~0.1 torr), whereupon a red-orange powder is obtained. After its dissolution in 20 mL of petroleum ether (45–65° fraction) two equivalents of trimethyl phosphite* are added under stirring at room temperature. The solution turns yellow and evolution of CO is observed. The stirring is maintained for 10 minutes, after which some crystals appear at room temperature. The tube is cooled to −25° and yellow crystals are obtained. The supernatant solution is transferred to another tube, concentrated under reduced pressure, and maintained at −25° for several days. The solid is dried under vacuum, and the overall yield is 90% (1.433 g). *Anal.* Calcd. for $C_{16}H_{36}O_8P_2Rh_2S_2$: C, 27.92; H. 5.27; P, 9.00; S, 9.31. Found: C, 28.11; H, 5.47; P, 9.12; S, 9.35. mp 99° (in air).

Properties

The $[Rh_2(\mu\text{-}t\text{-}C_4H_9S)_2(CO)_2\{P(OCH_3)_3\}_2]$ complex obtained by this procedure is a yellow, crystalline material possessing high solubility in many organic solvents. The complex in the solid state is slightly sensitive to oxygen, and solutions of it are rapidly decomposed upon exposure to air. It is advisable to store the compound under nitrogen to maintain its purity. It is a good catalyst for hydrogenation reactions, especially for the hydroformylation of alkenes.[8]

References

1. (a) J. Chatt and L. M. Venanzi, *J. Chem. Soc.*, 4735 (1957); (b) G. Giordano and R. H. Crabtree, *Inorg. Synth.*, **19,** 218 (1979); but in this case the preparation was performed in the absence of sodium carbonate.
2. All the manipulations were carried out under an efficient fume hood on a vacuum line by use of the classical Schlenk tube technique.[3] All the solvents were dried and stored under a nitrogen atmosphere.
3. D. F. Shriver, *The Manipulation of Air-Sensitive Compounds*, McGraw-Hill, New York, 1969.
4. L. M. Haines, *Inorg. Chem.*, **9,** 1517 (1970).
5. M. L. Wu, M. J. Desmond and R. S. Drago, *Inorg. Chem.*, **18,** 679 (1979).
6. D. de Montauzon and R. Poilblanc, *Inorg. Synth.*, **20,** 237 (1980).
7. Ph. Kalck, T. Poilblanc, R. P. Martin, A. Rovera, and A. Gaset, *J. Organomet. Chem.*, **195,** C 9 (1980).
8. Ph. Kalck, R. Poilblanc, and A. Gaset, U. S. Pat. 421,5066 (1979), *C.A.*, **91,** 129,602 f (1979).
9. Ph. Kalck and R. Poilblanc, *Inorg. Chem.*, **14,** 2779 (1975).
10. J. A. McCleverty and G. Wilkinson, *Inorg. Synth.*, **8,** 211 (1966) (prepared form $RhCl_3 \cdot 3H_2O$†).

*Available from Aldrich Chemical Co., Milwaukee, WI.
†Available from Alfa Products, Ventron Corp., Danvers, MA.

11. It is possible to check (by infrared techniques) that the reaction has occurred completely:
$[Rh_2(\mu\text{-}Cl)_2(CO)_4]$ $\nu CO(cm^{-1})$: 2105 m, 2089 vs, 2034 vs
$[Rh_2(\mu\text{-}Cl)(\mu\text{-}t\text{-}C_4H_9S)(CO)_4]$: 2088 m, 2072 vs, 2014 vs
$[Rh_2(\mu\text{-}t\text{-}C_4H_9S)_2(CO)_4]$: 2067 m, 2047 vs, 1998 vs

25. DINUCLEAR METHOXY, CYCLOOCTADIENE, AND BARRELENE** COMPLEXES OF RHODIUM(I) AND IRIDIUM(I)

Submitted by R. USON,* L. A. ORO,* and J. A. CABEZA*
Checked by H. E. BRYNDZA† and M. P. STEPRO†

Bis(η^4-cyclooctadiene)-di-μ-methoxy dirhodium(I), $[Rh(OMe)(1,5\text{-cod})]_2$, has been prepared by the reaction of $[RhCl(1,5\text{-cod})]_2$ with sodium carbonate in refluxing methanol,[1] or $[Rh(1,5\text{-cod})(indole)]ClO_4$ with potassium hydroxide in methanol.[2] The latter method has also been used for the synthesis of analogous derivatives containing tetrafluorobenzobarrelene,‡ $[Rh(OMe)(tfb)]_2$, and trimethyltetrafluorobenzobarrelene $[Rh(OMe)(Me_3tfb)]_2$.[2] The dioxygen-bridged complex $[Rh_2(O_2)(1,5\text{-cod})_2]$ reacts with methanol or water to give $[Rh(OMe)(1,5\text{-cod})]_2$ or the hydroxo-bridged complex $[Rh(OH)(1,5\text{-cod})]_2$.[3]

Bis(η^4-cyclooctadiene)-di-μ-methoxy diiridium(I), $[Ir(OMe)(1,5\text{-cod})]_2$, has been synthetized from three different starting materials, $[IrCl(1,5\text{-cod})]_2$,[4] $[IrHCl_2(1,5\text{-cod})]_2$,[5] or $[Ir(1,5\text{-cod})_2][BF_4]$[6] by allowing them to react with sodium carbonate in methanol.

The methoxide ion is a strong base, capable of taking a proton from any acid stronger than methanol. The methoxy-bridged complexes have been used both for synthetic purposes[7] and as homogeneous catalyst precursors.[8]

If potassium hydroxide is substituted for sodium carbonate in the preparation of methoxy-bridged derivatives from chloro-bridged complexes, the reactions take place readily under mild conditions, and lead to excellent yields. The use of water as a precipitating agent and as a washing fluid is of great importance, and its use in previously-described processes increases the yields. The syntheses of several methoxy-bridged complexes of rhodium(I) or iridium(I), along with

*Department of Inorganic Chemistry, University of Saragossa, Saragossa, Spain.

**Barrelene = bicyclo[2.2.2.]octa-2,5,7-triene.

†Central Research and Development Department, E. I. du Pont de Nemours and Co., Wilmington, DE 19898.

‡Tetrafluorobenzobarrelene = 5,6,7,8-tetrafluoro-1,4-dihydro-1,4-ethenonaphthalene.

the preparation of bis(η^4-1,5-cyclooctadiene)-di-μ-hydroxo-dirhodium(I) are described herein.

A. BIS(η^4-1,5-CYCLOOCTADIENE)-DI-μ-METHOXY-DIRHODIUM(I)

$$[RhCl(1,5\text{-cod})]_2 + 2\ KOH + 2\ MeOH \longrightarrow [Rh(OMe)(1,5\text{-cod})]_2 + 2\ KCl + 2\ H_2O$$

Procedure

A 50-mL round-bottomed flask containing a magnetic stirring bar is charged with a solution of $[RhCl(1,5\text{-cod})]_2$[9] (175 mg, 0.355 mmole) in dichloromethane (15 mL). The addition of a solution of potassium hydroxide (40 mg, 0.713 mmole) in methanol (5 mL) gives rise to the immediate precipitation of a yellow solid. After being stirred for 30 minutes at room temperature, the solvent is completely removed in a rotary evaporator. Then, 10 mL of methanol and subsequently 15 mL of water are added to the residue, after which the solid is collected by filtration using a fine sintered-glass filter, washed with water (ten 5-mL portions) and vacuum-dried over phosphorus(V) oxide. Yield: 157 mg (92%). The crude product can be used without further purification. An analytically pure sample can be obtained by recrystallization from a mixture of dichloromethane and hexane. *Anal.* Calcd. for $C_{18}H_{30}O_2Rh_2$: C, 44.65; H, 6.24. Found: C, 44.90; H, 6.20.

Properties

The yellow, air-stable compound is soluble in dichloromethane and chloroform, slightly soluble in benzene and acetone, less soluble in diethyl ether and hexane, and insoluble in water. On being heated, it decomposes between 120–175°. Its ^1H nmr spectrum ($CDCl_3$, TMS) shows three, broad multiplets at δ3.55 (8H, vinylic protons), 2.47 (8H, allylic protons) and 1.63 (8H, allylic protons), along with a sharp singlet at δ2.67 (6H, methoxide). IR (Nujol, cm^{-1}): 3328(m, br), 1330(m), 1304(m), 1231(w), 1214(w), 1154(w), 1078(vs), 998(m), 953(s), 892(w), 865(m), 815(m), 797(w), 775(m), 553(s), 500(m), 338(m).

Analogous Complexes

The tetrafluorobenzobarrelene and trimethyltetrafluorobenzobarrelene complexes, $[Rh(OMe)(tfb)]_2$ or $[Rh(OMe)(Me_3tfb)]_2$, can be prepared by the same method using $[RhCl(tfb)]_2$[10] and $[RhCl(Me_3tfb)]$,[10] respectively, as starting compounds.

For $[Rh(OMe)(tfb)]_2$: Yield: 94%. *Anal.* Calcd. for $C_{26}H_{18}F_8O_2Rh_2$: C, 43.36; H, 2.52. Found: C, 43.76; H, 2.37. The yellow, air-stable compound is slightly soluble in organic solvents and insoluble in water. It decomposes at 215–220°. IR (Nujol, cm^{-1}): 1649(w), 1489(vs), 1438(s), 1317(s), 1303(s), 1251(w), 1176(s), 1098(s), 1073(m), 1055(vs), 1038(vs), 947(s), 931(w), 914(m), 897(s), 859(s), 844(m), 685(m), 656(w), 489(s), 450(w), 430(s), 416(m), 382(m), 338(m), 290(m).

For $[Rh(OMe)(Me_3tfb)]_2$: Yield 87%. *Anal.* Calcd. for $C_{32}H_{30}F_8O_2Rh_2$: C, 44.80; H, 3.76. Found: C, 45.02; H, 3.85. The yellow, air-stable compound is soluble in dichloromethane and chloroform, slightly soluble in methanol, acetone, hexane, and diethyl ether, and is insoluble in water. It decomposes at 175–185°. ^1H nmr, ($CDCl_3$, TMS, δ): 4.91(m, 2H, ⋗CH), 2.97(m, 4H, =CH—), 2.48(br, 12H, ⋗C—Me, OMe), 1.47(s, br, 12H, ⋗C—Me). IR (Nujol, cm^{-1}): 1639(w, br), 1499(s), 1492(s), 1368(m), 1309(m), 1290(m), 1251(m), 1216(w), 1159(m), 1114(m), 1093(m), 1070(vs), 1056(vs), 1043(sh), 1032(sh), 1002(w), 974(s), 929(s), 910(s), 873(w), 852(s), 769(m), 685(m), 653(w), 579(m), 493(m), 370(w), 342(w).

B. BIS(η^4-1,5-CYCLOOCTADIENE)-DI-μ-METHOXY-DIIRIDIUM(I).

$$[IrCl(1,5\text{-cod})]_2 + 2\,KOH + 2\,MeOH \longrightarrow [Ir(OMe)(1,5\text{-cod})]_2 + 2\,KCl + 2\,H_2O$$

Procedure

Note. This preparation should be performed with deoxygenated solvents and in an inert atmosphere.

A 100-ml Schlenk flask containing a magnetic stirring bar, and equipped with an argon inlet, is charged with a solution of potassium hydroxide (25 mg, 0.445 mmole) in methanol (5 mL). Addition of a suspension of $[IrCl(1,5\text{-cod})]_2$[11] (149 mg, 0.222 mmole) in methanol (10 mL) turns the color of the initially orange-red suspension to yellow. After being stirred for 30 minutes at room temperature, 40 mL of water are added. The yellow precipitate is collected by filtration, using a fine sintered-glass filter, washed with water (six 5-mL portions), and vacuum-dried over phosphorus(V) oxide. Yield: 124 mg (85%).* *Anal.* Calcd. for $C_{18}H_{30}Ir_2O_2$: C, 32.62; H, 4.56. Found: C, 32.54; H, 4.38.

*A yield of 75% was reported by the checkers.

Properties

The yellow, air-stable compound is soluble in chlorinated solvents, and the resulting solutions are air-sensitive. The compound is slightly soluble in methanol, acetone, hexane, benzene, and diethyl ether, and is insoluble in water. It decomposes at 145–165°. The ^1H nmr spectrum ($CDCl_3$, TMS) shows three broad multiplets at δ3.57 (8H, vinylic protons), 2.22 (8H, allylic protons) and 1.45 (8H, allylic protons), and a sharp singlet at δ3.28 (6H, methoxide). IR (Nujol, cm^{-1}): 1325(m), 1300(m), 1232(w), 1208(m), 1172(w), 1158(w), 1060(vs), 1005(m), 972(s), 913(m), 894(w), 827(m), 811(w), 783(w), 574(s), 560(sh), 532(m), 512(w), 434(m), 337(m). An X-ray crystallographic study of this complex has been reported.[12]

C. BIS(η^4-1,5-CYCLOOCTADIENE)-DI-μ-HYDROXO-DIRHODIUM(I)

$$[RhCl(1,5\text{-cod})]_2 + 2\ KOH \longrightarrow [Rh(OH)(1,5\text{-cod})]_2 + 2\ KCl$$

Procedure

To a 50-mL round-bottomed flask containing a magnetic stirring bar and charged with a solution of potassium hydroxide (75 mg, 1.337 mmole) in water (4 mL), $[RhCl(1,5\text{-cod})]_2$[9] (320.5 mg, 0.65 mmole) in acetone (35 mL) is added. After being stirred for two hours at room temperature, the yellow suspension is concentrated to ~10 mL in a rotary evaporator. Then 15 mL of water is added. The solid is collected by filtration, using a fine sintered-glass filter, washed with water (ten 5-mL portions), and vacuum-dried over phosphorus(V) oxide. Yield: 280 mg (95%). It can be recrystallized from a mixture of dichloromethane and hexane. *Anal.* Calcd. for $C_{16}H_{26}O_2Rh_2$: C, 42.13; H, 5.74. Found: C, 42.23; H, 5.47.

Properties

The pale-yellow, air-stable solid is soluble in chlorinated solvents, slightly soluble in acetone, hexane, and diethyl ether, and is insoluble in water. It decomposes at 138–145°. MW: Calcd.: 456; Found: 484 (osmometrically in chloroform at 35°). The ^1H nmr spectrum ($CDCl_3$, TMS) shows a broad singlet at δ3.82 (8H, vinylic protons) and two broad multiplets at δ2.45 (10H, allylic protons and hydroxide) and 1.67 (8H, allylic protons). The IR (Nujol, cm^{-1}): 3592(s), 3390(m, br). 3300(m,br), 3190(m, br), 1327(m), 1300(w), 1230(w), 1212(m), 1176(w), 1156(w), 1080(w), 998(m), 962(s), 873(sh), 868(m), 817(m), 771(w),

438(m, br), 456(w), 440(w), 397(m), 359(w), 331(w). The ν(OD) in a deuterated sample*: 2628(s), 2501(m, br), 2416(m, br), 2357(m, br).

References

1. J. Chatt and L. M. Venanzi, *J. Chem. Soc.*, 4735 (1957).
2. R. Usón, L. A. Oro, J. A. Cabeza, and M. Valderrama, *J. Organometal. Chem.*, **231,** C81 (1982).
3. F. Sakurai, H. Suzuki, Y. Moro-oka and T. Ikausa, *J. Am. Chem. Soc.*, **102,** 1749 (1980).
4. G. Pannetier, P. Fougeroux, R. Bonnaire, and N. Platzer, *J. Less-Common Metals*, **24,** 83 (1971).
5. S. D. Robinson and B. L. Shaw, *J. Chem. Soc.*, 4997 (1965).
6. M. Green, T. A. Kuc and S. H. Taylor, *J. Chem. Soc.* (*A*), 2334 (1971).
7. (a) N. Platzer, N. Goasdone, and R. Bonnaire, *J. Organomet. Chem.*, **160,** 455 (1978); (b) S. W. Kaiser, R. B. Saillant, W. M. Butler, and P. G. Rasmussen, *Inorg. Chem.*, **15,** 2681 (1976); (c) B. Denise and G. Pannetier, *J. Organomet. Chem.*, **63,** 423 (1973); (d) R. Usón, L. A. Oro, and J. A. Cuchi, *Rev. Acad. Ciencias Zaragoza*, **31,** 173 (1975); (e) A. C. Sievert and E. L. Muetterties, *Inorg. Chem.*, **20,** 489 (1981).
8. (a) A. Krzywicki and G. Pannetier, *Bull. Soc. Chim. Fr.*, 1093 (1975); (b) D. Brodzki, C. Leclere, B. Denise, and G. Pannetier, *Bull. Soc. Chim. Fr.*, 61 (1976); (c) D. Brodzki, B. Denise, and G. Pannetier, *J. Mol. Catal.*, **2,** 149 (1977); (d) D. Brodzki, B. Denise, R. Bonnaire, G. Pannetier, and C. Leclere, *Fr. Demande*, **2,** 317, 269 (1977); (e) V. A. Kormer, B. D. Babitskii, M. I. Lovach, and N. N. Chesnokova, J. *Polymer, Sci.*, **C16,** 4351 (1969).
9. G. Giordano and R. H. Crabtree, *Inorg. Synth.*, **19,** 218 (1979).
10. D. M. Roe and A. G. Massey, *J. Organomet. Chem.*, **28,** 273 (1971).
11. J. L. Herde, J. C. Lambert, and C. V. Senoff, *Inorg. Syn.*, **15,** 18 (1974).
12. D. Tabrizi and G. Pannetier, *J. Less-Common. Metals*, **23,** 443 (1971).

26. AQUA IONS OF MOLYBDENUM

Submitted by D. T. RICHENS† and A. G. SYKES†
Checked by Z. DORI‡

Molybdenum is at present unique in having aqua ions in five oxidation states. Whereas the complexities of Mo(VI) aqueous solution chemistry have been understood in general terms for some time, it is only in the last 15 years that the aqua ions of the lower oxidation states II through V have been identified, and their structures clearly established. Metal aqua ions are notoriously difficult to crystallize for X-ray diffraction studies, and structures of derivative complexes

*Prepared by substituting deuterium oxide for water.

†Department of Inorganic Chemistry, The University, Newcastle-upon-Tyne, NE1 7RU, United Kingdom.

‡Department of Chemistry, Technion-Israel Institute of Technology, Haifa, Israel.

have to be determined instead. The species $[Mo(H_2O)_6]^{3+}$ is the only monomeric hexaaqua ion established. Multiple metal-metal bonds, mixed-valence compounds, triangular clusters and polymeric species (features which are often regarded as characteristic of other areas of chemistry) are all present. Table I summarizes known ions, the preparations of which are described herein, to which can be added the recently prepared trimeric Mo(III) aqua ion. Syntheses of molybdenum in oxidation states (II) and (III) require rigorous O_2-free techniques using N_2 (or Ar) gas, rubber seals (i.e., serum caps from medical supplies), syringes, and Teflon tubing and/or stainless steel needles (20G, i.e., 0.6 mm diameter). The IV and V state ions must also be stored air-free, although oxidation is not extensive over ~one-hour periods. Perchlorate cannot be used with the II and III ions, and also appears to oxidize the IV state, although in a somewhat random manner. Instead, non-complexing, redox-inactive, and strongly acidic trifluoromethanesulfonic acid, CF_3SO_3H (abbreviated HTFMS) and *p*-toluenesulfonic acid, $C_6H_4(CH_3)(SO_3H)$ (abbreviated HPTS), both of which can be obtained commercially, are used. The former is purified by redistillation and the latter by recrystallization.

A. MOLYBDENUM(II) AQUA DIMER:

$$[Mo(CO)_6] \longrightarrow [Mo_2(O_2CCH_3)_4] \longrightarrow [Mo_2Cl_8]^{4-} \longrightarrow [Mo_2(SO_4)_4]^{4-} \longrightarrow [Mo_2(H_2O)_8]^{4+}$$

Commercially available hexacarbonyl molybdenum is used as the starting material. The procedure involves four stages, requiring the preparation of tetra-μ-acetato-dimolybdate(II), $[Mo_2(O_2CCH_3)_4]$,[2] the octachlorodimolybdate(II) complex, $K_4[Mo_2Cl_8]\cdot 2H_2O$,[3] and the tetra-μ-sulfato-dimolybdate(II) complex, $K_4[Mo_2(SO_4)_4]\cdot 3.5H_2O$. The molybdenum(II) aqua ion is generated in the final step by sulfate abstraction, using Ba^{2+} in trifluoromethanesulfonic acid solution.[5]

Procedure

The tetra-μ-acetato-dimolybdenum(II) can be prepared from hexacarbonyl molybdenum by a method already described.[2] The yellow crystals obtained are used for the second stage without further purification.

Conversion to the tetrapotassium octachlorodimolybdate(II) dihydrate[3] is achieved as follows. Concentrated hydrochloric acid (130 mL), in a 250-mL conical flask, is cooled to 0° in ice, and saturated with hydrogen chloride gas for seven minutes. Potassium chloride (3.6 g) is added with stirring, and the mixture warmed to room temperature. The tetra-μ-acetato-dimolybdate(II) (2.6 g) is added, and the mixture is stirred for one hour. The purple-red precipitate is filtered through a fine sintered-glass funnel, washed with absolute ethanol (20 mL), and dried under a stream of N_2 gas. Final drying is in a vacuum desiccator (10^{-2} torr) over potassium hydroxide pellets for six hours. The yield of tetrapotassium octachlorodimolybdate(II) dihydrate is 3.3 g (85%). The third stage is carried out in an O_2-free atmosphere. A solution of tetrapotassium octachlorodimolybdate(II) dihydrate (3.3 g) in O_2-free 0.1 *M* sulfuric acid (300 mL) is stirred for three hours at room temperature in a 500-mL conical flask. The flask is then transferred to a glove box or bag filled with N_2, and potassium sulfate (15 g) is added. The mixture is stirred for an additional 10 minutes. The resulting pink precipitate is filtered through a fine sintered-glass funnel, washed with absolute ethanol (two × 30 mL) and diethyl ether (30 mL), and finally dried to constant weight in a vacuum desiccator (10^{-2} torr) over silica gel. The yield is 3.6 g (95%) of tetrapotassium tetra-μ-sulfato-dimolybdate(II). The material, when stored at 0–4° in a refrigerator under N_2, appears to be stable indefinitely.

The final stage requires rigorous O_2-free conditions. Tetrapotassium tetra-μ-sulfato-dimolybdate(II) is dissolved in an O_2-free solution of trifluoromethane sulfonic acid (redistilled at ~15 torr) in a ~100 mL capacity cylindrical centrifuge tube fitted with a rubber seal, and barium trifluoromethane sulfonate, prepared by addition of barium hydroxide to the acid ($Ba^{2+} \leqslant 20\%$ of the excess required to produce the sulfate), is added. The total volume should be about 50 mL, with the trifluoromethane sulfonic acid at ~0.1 *M*. The solution is carefully heated to 40° for five minutes (to help coagulation), and then cooled to 0° in an ice-bath to complete the precipitation of barium sulfate. As a precaution, N_2 is passed through the tube by way of stainless steel or Teflon entry and exit leads during this procedure. The barium sulfate is finally removed by centrifugation, and the pink solution transferred using Teflon tubing and N_2 gas pressure. Solutions under N_2 can be kept for three-hour periods. Stoichiometric quantities of Ba^{2+} can be used with more time and care for the precipitation.

Properties

The diamagnetic molybdenum(II) aqua ion is characterized by an absorption maximum at 510 mm ($\epsilon = 370\ M^{-1}\ cm^{-1}$) ($\delta \rightarrow \delta^*$ transition), with a minimum at 420 mm ($\epsilon = 120\ M^{-1}\ cm^{-1}$) for the dimer. Pure samples do not have a peak at 370 nm. The oxidation state of the molybdenum may be verified as II by addition of a 100-fold excess of Fe(III), to convert Mo(II) to Mo(VI) (20 minutes required), and titration of the generated Fe(II) with standard Ce(IV), using ferroin as the indicator.

A structure containing two quadruply bonded molybdenum(II) atoms surrounded by eight water molecules in an "eclipsed" configuration can be assigned, in view of the similarity in absorption spectrum of the aqua ion, the tetra-μ-sulfato-dimolybdate(II) ion, (λ_{max} 516 nm), and the octachlorodimolybdate(II) ion, (λ_{max} 515 nm). The crystal structure of these two complexes have

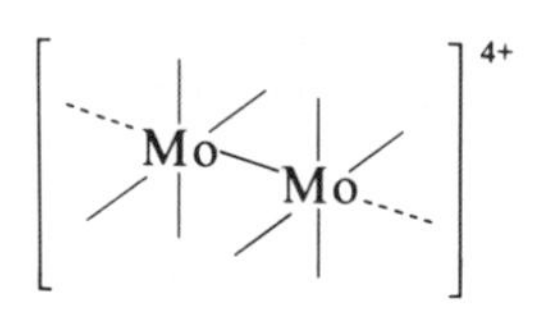

been reported, and the Mo–Mo distances are ~2.1Å. Two additional weakly bonded, axial water molecules are likely.

The molybdenum(II) aqua dimer can be oxidized electrochemically in 1 *M* hydrochloric acid to yield the unstable triply-bonded tetrachlorodimolybdenum(III) cation, $[Mo_2Cl_4]^{2+}$ (λ_{max} 430 nm).[6] A triply-bonded dimeric Mo(III) complex is also obtained by allowing a solution of $K_4[Mo_2Cl_8] \cdot 2H_2O$ in aqueous orthophosphoric acid (2 *M*) to stand in air for 24 hours. The addition of Cs^+ allows purple crystals of $Cs_2[Mo_2(HPO_4)_4(H_2O)_2]$, with axial H_2O's, to be obtained.[7] Oxidation of a saturated solution of $K_4[Mo_2(SO_4)_4]\cdot 3.5H_2O$ in 2 *M* sulfuric acid by an air stream yields the blue Mo(II,III) μ-sulfato analogue, the crystal structure of which has been determined. Irradiation of Mo_2^{4+} in 1 *M* HTFMS at 254 nm gives the green molybdenum(III) aqua dimer (see later) and H_2 gas.[9]

B. HEXAAQUAMOLYBDENUM(III)

$$[MoCl_6]^{3-} + 6H^+ + 6H_2O \longrightarrow [Mo(H_2O)_6]^{3+} + 6HCl$$

The starting materials, ammonium hexachloromolybdate(III), $(NH_4)_3[MoCl_6]$, $K_3[MoCl_6]$, or ammonium aquapentachloromolybdate(III), $(NH_4)_2[MoCl_5(H_2O)]$, can be prepared as stable solids according to literature methods.[10] The following procedure is then carried out under rigorous O_2-free conditions with a brisk stream of N_2 gas.

Procedure

Either $(NH_4)_3[MoCl_6]$, $(NH_4)_2[MoCl_5(H_2O)]$, or $K_3[MoCl_6]$, (1.2 g) is dissolved in 200 mL of O_2-free 0.5 *M* HPTS, and kept at room temperature for 24 hours. Over longer periods, other Mo(III) species can be formed.[11] The solution is diluted to 0.3 *M* (in HPTS) and loaded onto a 16 cm × 1 cm column of Dowex 50W-X cation-exchange resin (H^+ form), previously washed with O_2-free water

(250 mL). The column is jacketed with running ice-cold water, and the solution maintained O_2-free by a procedure such as that indicated in the diagram. A

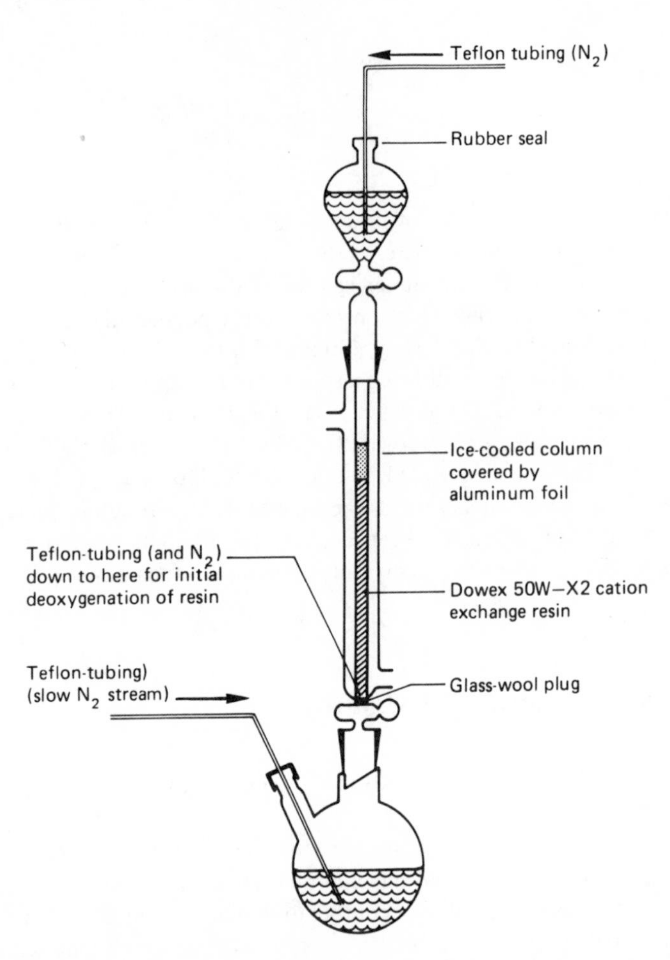

greenish-yellow band is obtained on the column, with a smaller red-brown band at the top of the column. The column is washed with 0.2 *M* HPTS (100 mL), followed by 1.0 *M* HPTS (100 mL), which elutes small amounts of a yellow component, probably molybdenum(V). With 2.0 *M* HPTS (100–200 mL), a pale-yellow solution of $[Mo(H_2O)_6]^{3+}$ is obtained. Solutions are best transferred using narrow gauge Teflon tubing and N_2 gas pressure. Stock solutions of $[Mo(H_2O)_6]^{3+}$ (concentrations up to 0.060 *M* have been prepared) in 1 *M* HPTS can be stored at 0–4° in a refrigerator under a N_2 atmosphere for at least one week without serious deterioration. A convenient method of storage is to place a 60-mL sample container filled with N_2 gas inside a larger 200-mL container, also filled with N_2, thus providing double protection against atmospheric oxidation.

Properties

Solutions of $[Mo(H_2O)_6]^{3+}$ in 1–2 *M* HPTS give d-d bands at 310 nm (ϵ = 23.2 M^{-1} cm^{-1}) and 380 nm (ϵ = 13.8 M^{-1} cm^{-1}). The shift of bands into the UV as compared to those for $[Cr(H_2O)_6]^{3+}$ at 408 nm and 587 nm is noted. The strong absorbance below 300 nm is due to HPTS. The concentration of Mo^{3+} can be determined by oxidation as described for the Mo(II) aqua ion.

Reactions with hexachloroiridate(IV) (fast), O_2 ($t_{1/2}$ ~ 1 min), and aqua VO^{2+} ($t_{1/2}$ ~ 10 min in a typical experiment) yield the molybdenum(V) aqua dimer as the only identifiable product. At much slower rates, O_2 and hexachloroiridate(IV) also oxidize Mo(V) to Mo(VI).

C. MOLYBDENUM(III) AQUA DIMER

$$2[MoO_4]^{2-} + 14\,H^+ + 3\,Zn \longrightarrow [Mo_2(OH)_2]^{4+} + 3\,Zn^{2+} + 6\,H_2O.$$

Procedure

This ion is obtained by reduction of sodium molybdate, $Na_2[MoO_4]\cdot 2H_2O$ (0.01 *M*, 100 mL), in 1.0 *M* HPTS using a Jones reductor (Zn/Hg) column under rigorously O_2-free conditions.[12,13] The Zn/Hg is prepared by stirring zinc shot (8–30 mesh) in a 2% solution of mercury(II) chloride for 10 minutes. After washing three times with 2 *M* hydrochloric acid, a column (25 cm × 1 cm) can be assembled. The column is initially washed with O_2-free water (250 mL) to exclude all gas present. The Mo(VI) is then passed down the column and the solution obtained is transferred through Teflon tubing using N_2 gas pressure. After dilution of the HPTS to 0.5 *M*, the blue-green solution obtained is loaded onto an ice-cold 30 cm × 1 cm column of Dowex 50W − X2 cation-exchange resin (H^+ form). A sharp, blue-green band is held on the column, and lighter green material is less strongly held. On washing first with 0.5 *M* HPTS (100 mL) and then 1 *M* HPTS (100 mL), a pale, yellow-brown solution containing Mo(V) is obtained. The blue-green molybdenum(III) aqua dimer is finally eluted with 2 *M* HPTS (200 mL).

Stock solutions of $Mo(III)_2$ (typically 10^{-2} *M*) can be stored at 0° in 2 *M* HPTS under N_2 for up to two weeks without serious deterioration (<5%).

Properties

The absorption spectrum in 2 *M* HPTS gives peak positions λ/nm (ϵ/M^{-1} cm^{-1}) at 360(910), 572(96), and 624(110) (ϵ's per dimer).[13] Possible structures include the di-μ-hydroxo ion, $[(H_2O)_4Mo(OH)_2Mo(H_2O)_4]^{4+}$, which is an analog of the $[Mo_2(OH)_2(O_2CCH_3)(edta)]^-$ complex,[14] and the μ-oxo ion, $[(H_2O)_5Mo(O)Mo(H_2O)_5]^{4+}$. Air-oxidation yields the molybdenum(V) aqua dimer. Oxidations have been observed to be faster by $\sim 10^2$ than corresponding reactions of Mo^{3+}.

D. MOLYBDENUM(IV) AQUA TRIMER

$$[MoO_4]^{2-} + 2[MoCl_6]^{3-} + 12\ H^+ \longrightarrow Mo_3O_4^{4+} + 12\ HCl$$

Procedure

A solution of $Na_2[MoO_4]\cdot 2H_2O$ in 2 *M* hydrochloric acid (0.3 *M*, 25 ml) is added to a solution of $(NH_4)_3[MoCl_6]$ or $(NH_4)_2[MoCl_5(H_2O)]$ in 2 *M* hydrochloric acid (0.3 *M*, 50 mL). Both solutions should be made O_2-free by bubbling N_2 gas through them for 30 minutes prior to mixing. The reaction mixture is kept for one hour at 80–90° under N_2, and the resulting solution is stored at 0–4° in a refrigerator. A 10–20 mL sample of this stock solution is diluted to 50 times its volume with 0.5 *M* HPTS, and allowed to stand for at least one day, to allow aquation of the coordinated chloride. To isolate the aqua ion, the solution is transferred to a 16 cm × 1 cm Dowex 50W − X2 cation-exchange column (H^+ form). A dark-red band is formed with a diffuse yellowish-brown band of molybdenum(V) aqua dimer below this. Elution of the Mo(V) is achieved with 0.5 *M* HPTS (200 mL). The red band moves slowly under these conditions, and, in some cases, splits into two bands, the first of which is eluted with 1 *M* HPTS (100 mL) and contains chloride. The second band, the molybdenum(IV) aqua ion is eluted with 2 *M* HPTS (100 mL). Solutions that are 0.03 *M* in Mo(IV) are readily obtained. To obtain aqua Mo(IV) in a perchlorate medium, the 0.5 *M* HPTS solution can be exchanged onto a short (1–2 cm) Dowex 50W–X2 column and eluted with 2 *M* perchloric acid, after first moving the band to the bottom of the column with 1 *M* perchloric acid.

Properties

Aqua Mo(IV) in 2 *M* HPTS or $HClO_4$ has an absorption maximum at 505 nm ($\epsilon = 63\ M^{-1}$ cm^{-1}) and a minimum at 437 nm ($\epsilon = 54\ M^{-1}$ cm^{-1}), ϵ's per Mo atom.[15] Crystal structures of oxalate,[16] thiocyanate,[17] and edta[18] complexes have indicated a triangular structure with an apical μ_3-oxo ligand. The complexes and aqua ion (solutions that are 0.1 *M* in the trimer have been obtained) are diamagnetic. Using ^{18}O isotope labeling, it has been confirmed that this same Mo_3O_4 core structure is retained for the aqua ion in solution.[19] One of the three coordinated water molecules on each Mo(IV) is more inert to substitution ($t_{1/2}$ ~ one hour at 0°) than the other two.

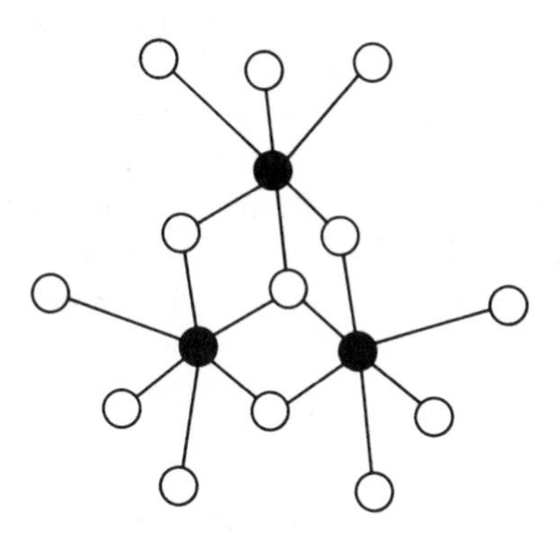

Solutions of aqua Mo(IV) in 2 *M* HPTS are stable to aerobic oxidation, and the loss of Mo(IV) is less than 10% per day. In the absence of O_2, such solutions appear to be stable indefinitely. The stability of Mo(IV) in O_2-free perchlorate solutions is variable, since solutions are often stable for more than one day at room temperature, but at other times for only ~one hour.[15]

Electrochemical reduction gives the pink-yellow Mo(III, III, IV) and green $Mo(III)_3$ aqua ions.[20] The latter can also be obtained by Zn/Hg reduction.

Aqua Mo(IV) monomer and dimer species have only transient existence as intermediates in the oxidation of the aqua Mo(III) and $Mo(III)_2$ ions. Strong oxidants, such as $IrCl_6^{2-}$ (0.89 v) and $Fe(phen)_3^{3+}$ (1.06 v), oxidize $Mo(IV)_3$ to $Mo(V)_2$ ($t_{1/2} > 1$ min), the slowness of the reaction reflecting the need for a structural change.

E. MOLYBDENUM(V) AQUA DIMER

$$4\ MoO_4^{2-} + N_2H_4 + 12\ H^+ \longrightarrow 2[Mo_2O_4]^{2+} + N_2 + 8\ H_2O.$$

This ion has been prepared by three different methods. These involve the reduction of sodium molybdate(VI) in 10 *M* hydrochloric acid with hydrazine hydrochloride at 80°,[21] the aquation of dipyridinium pentachlorooxomolybdate(V) in 0.05 *M* $HClO_4$,[22] and the reduction of sodium molybdate(VI) with ammonium hexachloromolybdate(III).[10] The first of these is described here.

Procedure

The Mo(V) oxidation state is generated by reduction of $Na_2[MoO_4] \cdot 2H_2O$ (8.2 g) in 10 *M* hydrochloric acid (180 mL) with hydrazine hydrochloride, (6.6 g) at 80° for two to three hours. The solution is concentrated about four-fold using a rotary evaporator. Portions of this green-brown solution (as required) can then be diluted 100-fold by the addition of O_2-free water to give a final $[H^+]$ of ~0.1 *M*. Under these conditions, the coordinated chloride ions aquate and the molybdenum(V) aqua dimer is formed. The solution is passed down a 12 cm × 1 cm column of Dowex 50W-X cation-exchange resin (H^+ form) under O_2-free conditions. The brown band of aqua $Mo(V)_2$ remaining on the column is washed with 0.2 *M* perchloric acid (100 mL) and eluted with 0.5–2 *M* perchloric acid, as required. Stock solutions of aqua $Mo(V)_2$ are typically 7×10^{-3} *M* in 0.5 *M* perchloric acid, but can be as high as 0.2 *M* when eluted with 2 *M* perchloric acid.

Properties

Solutions are standardized spectrophotometrically at 384 nm (ϵ = 103 M^{-1} cm^{-1}), and 295 nm (ϵ = 3,550 M^{-1} cm^{-1}), ϵ's per dimer.[21] The ion has a di-μ-oxo structure, formula: $[Mo_2O_4(H_2O)_6]^{2+}$. The crystal structure of tetrapyridinium di-μ-oxo-bis{trithiocyanatooxomolybdate(V)}, $(pyH)_4[Mo_2O_4(NCS)_6] \cdot H_2O$, indicates non-equivalent thiocyanate ligands.[23] Those Mo-NCS bonds *trans* to the terminal oxo ligand (2.30 Å) are longer than others (2.15 Å). The Mo-Mo distance is 2.58 Å. Similar structural features are to be expected for the aqua ion. The ethylenediaminetetraacetato complex has the edta in a basket-like configuration.[24] All dimeric forms are diamagnetic.

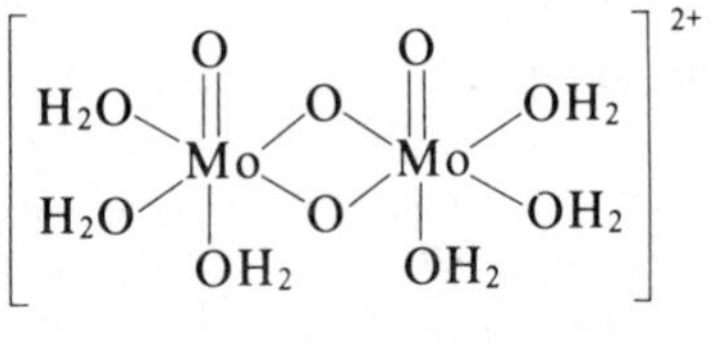

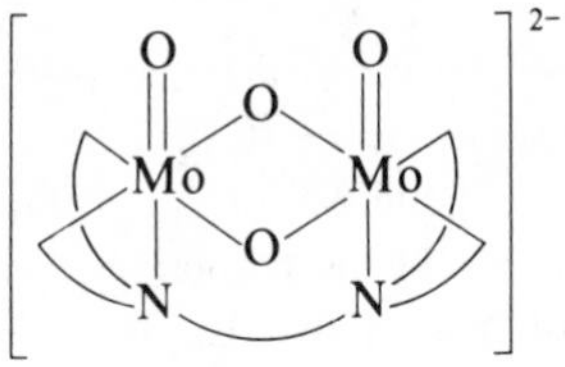

Solutions that are ~10^{-2} *M* in 0.5 *M* perchloric acid can be kept under N_2 for periods of at least two weeks. Variations in $HClO_4$ of 0.01–1 *M* have no measurable effect on the UV-visible spectrum, and a pK_a of > 2 is indicated. A yellow-to-orange color change observed between pH 2 and pH 3 is followed by precipitation. A polymeric aqua Mo(V) species, which has been reported,[25] yields $[Mo_2O_4(H_2O)_6]^{2+}$ quantitatively on acidification. The monomeric aqua Mo(V) ion generated electrochemically from Mo(VI) at high dilution (~10^{-5} *M*) in 2 *M* trifluoromethanesulfonic acid has been identified,[26] but undergoes rapid dimerization to $[Mo_2O_4(H_2O)_6]^{2+}$ (k ~ 10^3 M^{-1} sec^{-1}).

F. MOLYBDENUM(VI) AQUA IONS

Molybdenum(VI) is readily available commercially as sodium molybdate, $Na_2[MoO_4] \cdot 2H_2O$, and ammonium heptamolybdate, $(NH_4)_6[Mo_7O_{24}] \cdot H_2O$. At pH > 7 $[MoO_4]^{2-}$ is retained in solution. On decreasing the pH to <7, protonation induces a change in the coordination number of molybdenum from four to six, and polymerization to octahedral Mo_7 and Mo_8 species, $[Mo_7O_{24}]^{6-}$ and $[Mo_8O_{26}]^{4-}$, occurs. More acidic solutions ($[HClO_4]$ = 0.5–3.0 *M*) containing monomeric and dimeric species, have also been investigated.[27] Relevant equilibria can be expressed as follows,

$$2\,HMoO_3^+ \overset{K_1}{\rightleftharpoons} H_2Mo_2O_6^{2+}$$

$$HMo_2O_6^+ + H^+ \overset{K_2}{\rightleftharpoons} H_2Mo_2O_6^{2+}$$

$$H_2Mo_2O_6^{2+} + H^+ \overset{K_3}{\rightleftharpoons} H_3Mo_2O_6^{3+}$$

where the simplest possible formulae are indicated; $HMoO_3^+$ can also be written as, for example, $[Mo(OH)_5(H_2O)]^+$. At 25°, $K_1 = 97\ M^{-1}$, $K_2 = 4.7\ M^{-1}$, and $K_3 = 0.24\ M^{-1}$.[27]

Procedure

Solutions containing monomer and dimer aqua ions can be prepared from sodium molybdate(VI) by addition of dilute perchloric acid. As indicated above, the composition of solutions (0.3–9.0) × 10^{-3} *M* in Mo(VI), in 0.5 *M* $HClO_4$ (I = 3.0 *M* $NaClO_4$), has been specified.[27] Equilibration processes are known to be rapid.

Properties

All Mo(VI) aqua ions are colorless. Peaks are observed for the dimeric species in the 240–250 nm region.[27]

TABLE I Summary of Single Oxidation State Aqua Ions of Molybdenum

Description	Formula	Color
$Mo(II)_2$	$[Mo_2(H_2O)_8]^{4+}$	Red
Mo(III)	$[Mo(H_2O)_6]^{3+}$	Pale-yellow
$Mo(III)_2$	$[Mo_2(OH)_2(H_2O)_8]^{4+}$	Green
$Mo(IV)_3$	$[Mo_3O_4(H_2O)_9]^{4+}$	Red
$Mo(V)_2$	$[Mo_2O_4(H_2O)_6]^{2+}$	Yellow
$Mo(VI)_n$	Different forms	Colorless

References

1. D. T. Richens and A. G. Sykes, *Comm. Inorg. Chem.*, **1,** 141 (1981).
2. A. B. Brignole and F. A. Cotton, *Inorg. Syn.*, **13,** 88 (1972).
3. J. V. Brencic and F. A. Cotton, *Inorg. Chem.*, **9,** 351 (1970).
4. F. A. Cotton, F. A. Bertram, E. Pedersen, and T. R. Webb, *Inorg. Chem.*, **14,** 391 (1975).
5. A. R. Bowen and H. Taube, *J. Am. Chem. Soc.*, **93,** 3287 (1971), and *Inorg. Chem.*, **13,** 2245 (1974).
6. A. Bino, *Inorg. Chem.*, **20,** 623 (1981).
7. F. A. Cotton, B. A. Frenz, E. Pedersen, and T. R. Webb, *Inorg. Chem.*, **14,** 391 (1975).
8. A. Bino and F. A. Cotton, *Inorg. Chem.*, **18,** 3562 (1979).
9. W. C. Trogler, D. K. Erwin, G. L. Geoffroy, and H. B. Gray, *J. Am. Chem. Soc.*, **100,** 1160 (1978).
10. J. V. Brencic and F. A. Cotton, *Inorg. Syn.*, **13,** 170 (1972).
11. (a) Y. Sasaki and A. G. Sykes, *J. Chem. Soc. Chem. Comm.*, 767 (1973); (b) *J. Chem. Soc. Dalton, Trans.*, 1048 (1975).
12. M. Ardon and A. Pernick, *Inorg. Chem.*, **13,** 2276 (1974).
13. M. A. Harmer and A. G. Sykes, *Inorg. Chem.*, **20,** 3963 (1981).
14. G. G. Kneale, A. J. Geddes, Y. Sasaki, T. Shibahara, and A. G. Sykes, *J. Chem. Soc. Chem. Comm.*, 356 (1975).
15. M. A. Harmer, D. T. Richens, A. B. Soares, A. T. Thornton, and A. S. Sykes, *Inorg. Chem.*, **20,** 4155 (1981).
16. A. Bino, F. A. Cotton, and F. Dori, *J. Am. Chem. Soc.*, **100,** 5253 (1978).
17. E. O. Schlemper, M. S. Hussain, and R. K. Murmann, *Cryst. Struct. Comm.*, **11,** 89 (1982).
18. A. Bino, F. A. Cotton, and Z. Dori, *J. Am. Chem. Soc.*, **101,** 3842 (1979).
19. R. K. Murmann and M. E. Shelton, *J. Am. Chem. Soc.*, **102,** 3984 (1980).
20. D. T. Richens and A. G. Sykes, *Inorg. Chem.*, **21,** 418 (1982).
21. G. R. Cayley, R. S. Taylor, R. K. Wharton, and A. G. Sykes, *Inorg. Chem.*, **16,** 1377 (1977).
22. H. Sabat, M. F. Rudolf, and B. Jezowska-Trezbiatowska, *Inorg. Chim., Acta*, **7,** 365 (1973).
23. B. Jezowska-Trezbiatowska, *Russ. J. Inorg. Chem.*, **22,** 1950 (1977).
24. L. V. Haynes and D. T. Sawyer, *Inorg. Chem.*, **6,** 2146 (1967).
25. F. A. Armstrong and A. G. Sykes, *Polyhedron*, **1,** 109 (1982).
26. M. T. Paffett and F. C. Anson, *Inorg. Chem.*, **20,** 3967 (1981).
27. J. J. Cruywagen, J. B. B. Heyns, and E. F. C. H. Rohwer, *J. Inorg. Nucl. Chem.*, **40,** 53 (1978).

Chapter Five

UNUSUAL LIGANDS AND COMPOUNDS

27. 1,2-BIS(DICHLOROPHOSPHINO)ETHANE*

(Note: See **SPECIAL HAZARD NOTICE** in the Appendix of this volume.)

Submitted by RICHARD A. HENDERSON,† WASIF HUSSAIN,† G. JEFFERY LEIGH,† and FRED B. NORMANTON†
Checked by DAVID R. BRITTELLI and J. BURIAK, JR‡

$$8\ PCl_3 + P_4 + 6\ C_2H_4 \longrightarrow 6\ Cl_2PCH_2CH_2PCl_2$$

Introduction

Although the well-known and important ligand 1,2-ethanediylbis-(diphenylphosphine) ($Ph_2PCH_2CH_2PPh_2$) is conveniently prepared on a large scale in the laboratory,[1] the preparation of the analogous tetraalkyldiphosphines involves multi-step syntheses and toxic reagents.[2] However, a convenient and general preparation of tetraalkyl- and tetraaryl-diphosphines is afforded using 1,2-bis(dichlorophosphino)ethane* and the appropriate Grignard reagent.[3] The preparation of 1,2-bis(dichlorophosphino)ethane is based on the procedure patented by Toy and Uhing.[4]

*1,2-Ethanediylbis(dichlorophosphine).

†A.R.C. Unit of Nitrogen Fixation and School of Molecular Sciences, University of Sussex, Brighton, BN1 9RQ, United Kingdom.

‡Central Research and Development Department, E.I. du Pont de Nemours & Co., Wilmington, DE 19898.

Procedure

All reagents are used as supplied commercially. The autoclave and fittings (Baskerville and Lindsay 1-L heated rocking autoclave, constructed of Hastelloy) together with its glass liner are dried overnight, in an oven at 105°, prior to the preparation.*

The autoclave is loaded, in a well-ventilated fume-hood, with phosphorus trichloride (380 mL, 4.3 moles; Cambrian Chemicals). Yellow phosphorus (33.6 g, 1.08 moles; BDH Ltd.) is cut into small pieces (~1 mL in volume), dried rapidly with a tissue, and immediately added to the autoclave.

■ **Caution.** *Dry yellow phosphorus is spontaneously flammable in air, scrupulous drying of the phosphorus is not necessary, and gloves should always be worn when handling this material.*

The autoclave is then assembled and, in order to reduce the oxygen and hydrogen chloride to acceptable levels, it is twice filled with dinitrogen to 50 atm pressure and vented to a suitable fume-hood.

■ **Caution.** *During these, and all subsequent manipulations with the autoclave under pressure, ear protection should be worn in case the bursting disc blows.*

The autoclave is then charged once with ethylene to 10 atm and vented (to flush air out of the connections to the gas bottle) and then charged with ethylene (Air Products, standard grade) to 48 atm, and set rocking at 20 times per minute. The temperature is gradually raised to 200°, so that the difference between the temperatures inside and outside the autoclave does not exceed 20°. The temperature is then maintained at 200 ° for 19 hours.

After allowing the autoclave to cool to room temperature (about six hours, with assistance from a fan), the residual pressure (<5 atm) is vented, and the autoclave is then purged with dinitrogen as before. The autoclave is then opened in a well-ventilated fume-hood, and the strongly fuming, dark brown solution, which contains some orange solid, is poured from the glass liner into a 500-mL Schlenk flask, and is flushed with dinitrogen. Any liquid in the steel autoclave is also transferred into the Schlenk flask.

A bright-orange solid is deposited both on the bottom of the thermometer pocket and the walls of the autoclave. The thermometer pocket is washed in a stream of cold water and the autoclave is filled with crushed ice as soon as the product has been removed.

*The authors used a loose-fitting glass liner in the autoclave, but the checkers report that the reaction products sometimes seal the liner into the autoclave, preventing its removal. Presumably this problem arises if the liner fits too well. The checkers preferred to use no liner. Whereas Hastelloy is reported[5] to be inert to the materials used in this reaction, our experience is that the surface of a normal stainless steel autoclave shows visible signs of corrosion after some six reaction batches. Consequently, the condition of such an autoclave should be checked regularly.

■ **Caution.** *If the orange solid is allowed to dry in air, it may ignite spontaneously.*

Using standard Schlenk techniques,[6] any orange solid which has been transferred into the flask is removed by filtration (porosity 3 sinter). Unchanged phosphorus trichloride (~185 cm^3) is removed by distillation (bp = 76°/760 torr). All further low-boiling material is removed under vacuum at room temperature. Finally, the product is distilled at 66–68° and 0.05 torr. Yield = 140–160 g.

■ **Caution.** *If the autoclave reaction is performed at <200°, a lower yield (~30g) is obtained, and, upon distillation, a white, pyrophoric solid is deposited in the condenser.*

Cleaning procedure

1. The autoclave is cleaned using warm water and soap-impregnated wire-wool. The glass liner, if used, is cleaned by flushing with a stream of cold water.
2. The black, viscous residue remaining after the distillation is dissolved in an aqueous sodium carbonate solution and washed away in a stream of cold water.
3. Treatment of washing residues with sodium hypochlorite solution before disposal prevents the spread of unpleasant phosphine-like smells.

Properties

1,2-Bis(dichlorophosphino)ethane is a colorless liquid which slowly oxidizes in air, and is best stored under nitrogen. The compound is characterized by mass spectrometry (M^+ 232). The ^{31}P nmr spectrum is a singlet δ −50.07 ppm relative to trimethylphosphite (external standard). The ^{1}H nmr spectrum consists of an $A_2A_2'XX'$ spin-system pattern centered at δ 2.59 ppm relative to tetramethylsilane. Addition of 1,2-bis(dichlorophosphino)ethane to an excess of a Grignard reagent in diethyl ether yields the appropriate tetraalkyl- or tetraaryl-diphosphine.[3]

References

1. W. Hewertson and H. R. Watson, *J. Chem. Soc.*, 1490 (1962).
2. S. A. Butler and J. Chatt, *Inorg. Syn.*, **15,** 185 (1974) and references therein.
3. R. J. Burt, J. Chatt, W. Hussain, and G. J. Leigh, *J. Organomet. Chem.*, **182,** 203 (1979).
4. A. D. F. Toy and E. H. Uhing, *U.S. Patent* No. 3,976,690, 1975.
5. G. A. Nelson, *Corrosion Data Survey,* Shell Development Company, Emeryville, CA (1960).
6. (a) D. F. Shriver, *The Manipulation of Air-Sensitive Compounds,* McGraw-Hill, New York, 1969, p. 145; (b) J. P. Collman, N. W. Hoffman, and H. W. Hosking, *Inorg. Syn.*, **12,** 10 (1970).

28. METAL CHELATES OF 2,2,7-TRIMETHYL-3,5-OCTANEDIONE, H(tod)

Submitted by THOMAS J. WENZEL,* ERIC J. WILLIAMS,† and ROBERT E. SIEVERS†
Checked by S. C. CUMMINGS‡ and L.-S. TAN‡

Introduction

Metal chelates containing beta-diketonate ligands have been used for a variety of purposes, including gas chromatography of metals,[1] metal plating on surfaces,[2] solvent extraction of metals,[3–5] catalysts, and fuel additives.[6,7] The physical properties of these complexes that determine their usefulness in these areas are thermal stability, volatility, and solubility in nonpolar solvents.

Of all the reported synthetic procedures for beta-diketones (see Reference 8 and references therein), perhaps the most widely used is the base-promoted Claisen condensation of a methyl ketone and an ester. The synthesis described here involves the coupling of the ethyl ester of pivalic acid (2,2-dimethylpropanoic acid)§ to methyl isobutyl ketone (4-methyl-2-pentanone) using sodium hydride as the base to produce (following acidification) the novel ligand, 2,2,7-trimethyl-3,5-octanedione, hereafter referred to as H(tod). For the reaction, quite inexpensive starting materials are employed, and the yield of the ligand is generally 70–80%. The procedure requires one to two days of laboratory time.

■ **Caution.** *Since hydrogen gas is generated by the action of sodium hydride, a well-ventilated fume hood is required, as is the exclusion of all open flames near the work area.*

A. 2,2,7-TRIMETHYL-3,5-OCTANEDIONE, H(tod)

$$(CH_3)_3CCO_2C_2H_5 + CH_3COCH_2CH(CH_3)_2 \xrightarrow[\text{Toluene}]{\text{NaH},\Delta} \xrightarrow[H_2O]{HCl}$$

$$(CH_3)_3CCOCH_2COCH_2CH(CH_3)_2 + CH_3CH_2OH$$

*Department of Chemistry, Bates College, Lewiston, ME 04240.

†Department of Chemistry and Cooperative Institute for Research in Environmental Sciences, University of Colorado, Boulder, CO 80309.

‡Department of Chemistry, Wright State University, Dayton, OH 45435.

§Also called neopentanoic acid (Exxon Chemicals, Houston, TX).

Procedure

To a 1-L, three-necked, round-bottomed flask are added a magnetic stirring bar,* 67.2 g (1.4 moles) of sodium hydride,** and 350 mL of dry toluene.† The flask is immersed in a hot water bath atop a stirrer-hot plate* and then fitted with a reflux condenser, a pressure-equalizing dropping funnel of at least 100-mL capacity, and a glass inlet for a nitrogen‡ purge gas system. Ethyl pivalate§ (106.5 mL, 0.70 mole) is added to the flask through one of the necks, and methyl isobutyl ketone¶ (88 mL, 0.70 mole) is added to the dropping funnel. Heat is applied to the water bath, stirring* is begun, and the purge gas is passed through the flask at a slow flow rate. When the water bath boils, the ketone is added at a rate of 0.5–1.0 mL/min, such that the entire volume is dispensed over the course of about three hours. When the addition is complete, the now green-yellow mixture is heated for an additional 30 minutes and then allowed to stir overnight without heat but with the purge gas flowing.

The next day the flask is packed in an ice bath, and, with the purge gas flowing, 300 mL of an 18% (6 *M*) aqueous hydrochloric acid solution is *slowly* added (dropwise, at first) with vigorous stirring (caution: *hydrogen evolution*). When the neutralization is complete, the mixture is filtered to remove the suspended solids, which are discarded. The mixture is transferred to a separatory funnel, and the organic layer is isolated. Subsequently, the solvent is removed by rotary evaporation.

The crude H(tod) is purified first by vacuum distillation (1 torr, 60–90°) to separate the ligand from high boiling but organic-soluble impurities. Second, the copper complex is synthesized, as described below, to isolate the beta-diketone from any by-products that are formed by self-condensation of the ketone. Crude [Cu(tod)$_2$] (20 g) is placed in a 250-mL separatory funnel with 30 mL of hexane and 150 mL of 10% (1 *M*) aqueous sulfuric acid. The mixture is shaken until no blue color remains in the organic phase. The organic solution is isolated and the solvent is removed by rotary evaporation. Generally, the H(tod) product is 97–99% pure.

Properties

The H(tod) is a colorless liquid at room temperature and has a disagreeable odor. The proton magnetic resonance spectrum of this compound measured in chlo-

*The checkers recommend use of a mechanical stirrer to give improved yields.

**50% dispersion in mineral oil; from Alfa Div. of Ventron Corp., Danvers, MA.

†Dried over 4A molecular sieves; from Fisher Scientific, Fair Lawn, NJ.

‡Passed through a trap containing 13× molecular sieves from Linde Corp.

§Synthesize by condensing ethanol and neopentanoic acid or obtain from Aldrich Chemical Co., Milwaukee, WI; store over andydrous magnesium sulfate; distill prior to use.

¶Dried over 3A molecular sieves, from Fisher Scientific Company, Fair Lawn, NJ.

roform solution shows a doublet at 0.90 ppm, a singlet at 1.17 ppm, a complex multiplet at 2.20 ppm, and singlets at 5.63 and 16.00 ppm. The signals correspond, respectively, to isobutyl CH_3 protons, *tert*-butyl CH_3 protons, isobutyl CH_2-CH protons, a methine proton, and an enol proton. The nmr spectrum also shows that H(tod) is predominantly in the enol form at 30 ° in chloroform solution. The mass spectrum of H(tod) shows major signals at 184 a.m.u., which is the molecular ion, 127 a.m.u., which is the base peak arising from the loss of a butyl group, and 57 a.m.u., which corresponds to a C_4H_9 fragment. The infrared spectrum of a neat sample on NaCl plates shows absorbances at 3000, 1550–1650, 1470, 1360, 1290, 1220, 1130, 950–970, 890, 870 and 780 cm^{-1}. The H(tod) is soluble in almost all organic solvents, but is only slightly soluble in water.

B. BIS(2,2,7-TRIMETHYL-3,5-OCTANEDIONATO)COPPER(II), [Cu(tod)$_2$]

$$Cu(CH_3CO_2)_2{\cdot}H_2O + 2\ H(tod) \xrightarrow[H_2O]{Methanol} [Cu(tod)_2] + 2\ H[CH_3CO_2) + H_2O$$

Procedure

To 100 mL of crude, vacuum-distilled H(tod) in a 1-L filter flask containing a magnetic stirring bar is added 100 mL of methanol. To a separate beaker containing 200 mL of distilled water is added 40 g of copper(II) acetate hydrate.* This solution is stirred and heated to just below the boiling point, and, while hot, is filtered directly into the stirred methanolic H(tod) solution. The resulting mixture is allowed to stand for four hours, after which the solid that forms is collected by suction filtration and washed with cold (4°) methanol (two × 10 mL). Recrystallization from hexane gives a purple powder. *Anal.* Calcd. for $Cu(C_{11}H_{19}O_2)_2$: Cu, 14.77; C, 61.44; H, 8.91. Found: Cu, 14.47; C, 61.48; H, 8.89.

Properties

The [Cu(tod)$_2$] is an air-stable, crystalline solid that melts at 116–117°. This compound has been shown by X-ray crystallography in this laboratory to be monomeric and square-planar. The unit cell is monoclinic with dimensions a = 20.369 Å,[4] b = 10.37 1 Å,[2] c = 11.526 Å,[2] and β = 90.04°.[1] The Cu(tod)$_2$ molecule crystallizes in the *cis*-configuration and is not solvated (as is also indicated by the absence of discontinuities during thermogravimetric analysis).

*The checkers report that 175 mL of water is needed to dissolve most of a 20-g sample of copper(II) acetate monohydrate.

The $Cu(tod)_2$ is much more soluble in hexane than almost all previously reported copper beta-diketonates.[3] The logarithm of the partition coefficient of $Cu(tod)_2$ between 1-octanol and water (log P_{oct}) is 3.64 ± 0.20. This value, which is unusually large for a metal complex, is an indication of the expected bioavailability[9] of $[Cu(tod)_2]$. We have shown that a 50% (v/v) solution of H(tod) in kerosene is capable of extracting copper(II) ion from ammoniacal aqueous solutions in high yield, which may be useful for commercial recovery of copper from ore leachates.[5]

C. TETRAKIS(2,2,7-TRIMETHYL-3,5-OCTANEDIONATO)-CERIUM(IV), $[Ce(tod)_4]$

$$4\ Ce(NO_3)_3{\cdot}6H_2O + 16\ H(tod) + 12\ NaOH + O_2 \xrightarrow[H_2O]{\text{Methanol}} 4\ [Ce(tod)_4] + 12\ NaNO_3 + 38\ H_2O$$

Procedure

To a stirred solution of $Ce(NO_3)_3{\cdot}6H_2O$ (10 g, 0.023 mole) in 30 mL of methanol is added a solution of H(tod) (17 g, 0.092 mole) in 50 mL of methanol. While the solution is being stirred to provide a continuous oxygen supply, 4 *M* aqueous sodium hydroxide is slowly added until the pH is brought up to a value of 7. During the addition of base, the solution turns a dark, red color, and precipitate begins to form at a pH of ~6.4. The solution is decanted and the thick, red product is dissolved in 100 mL of hexane. This solution is filtered through a medium-porosity, sintered-glass funnel and the solvent is removed by rotary evaporation, leaving dark, red crystals of $[Ce(tod)_4]$. The yield is 19.5 g (97%).

In some cases, an oil is obtained rather than crystals. If this occurs, crystallization is achieved by dissolving the oil in 100 mL of hexane and adding this solution to 400 mL of 95% ethanol with stirring. This solution is allowed to stand undisturbed overnight, during which time crystals of $Ce(tod)_4$ form. These crystals are collected by suction filtration, dissolved in hexane, and filtered through a medium-porosity, sintered-glass funnel. Removal of the solvent by rotary evaporation leaves dark, red crystals of $[Ce(tod)_4]$. *Anal.* Calcd. for $[Ce(C_{11}H_{19}O_2)_4]$: C, 60.52; H, 8.66. Found: C, 60.05; H, 8.66.

Properties

The compound $[Ce(tod)_4]$ is an air-stable, crystalline solid that melts at 134–136°. The proton magnetic resonance spectrum of $Ce(tod)_4$ in chloroform solution shows resonances (0.85 ppm, 1.09 ppm, and 5.30 ppm) that are characteristic of the ligand. The mass spectrum of $[Ce(tod)_4]$ exhibits a molecular ion at 872 a.m.u. The infrared spectrum was obtained using a KBr pellet of the pure compound. Absorbances were observed at 2947, 2860, 1562, 1403, 1217, 1154, 975, and 770 cm^{-1}.

D. TRIS(2,2,7-TRIMETHYL-3,5-OCTANEDIONATO)-MANGANESE(III), $[Mn(tod)_3]$

$$4\ Mn(CH_3CO_2)_2{\cdot}4H_2O + 12\ H(tod) + 8\ NaOH + O_2 \xrightarrow[H_2O]{Methanol} 4\ [Mn(tod)_3] + 8\ NaCH_3CO_2 + 26\ H_2O$$

Procedure

To 100 mL of methanol is added $Mn(CH_3CO_2)_2{\cdot}4H_2O$ (7.7 g, 0.031 mole) with stirring. After the solution turns from light pink to dark brown (about 15 min), H(tod) (18.4 g, 0.10 mole) is added and the pH is slowly increased by addition of 4 *M* aqueous sodium hydroxide with stirring to provide a continuous supply of oxygen. When the pH reaches approximately 8, addition of base is halted and the solution is stirred for an additional 30 minutes. The thick precipitate that forms is collected by suction filtration, dissolved in 100 mL of hexane, and the mixture is filtered through a medium-porosity, sintered-glass funnel.* The solvent is removed from the filtrate by rotary evaporation and the dark oil solidifies on standing at room temperature. The yield of $[Mn(tod)_3]$ is 10.3 g (55%). Recrystallization from hot methanol results in brown crystals. *Anal.* Calcd. for $[Mn(C_{11}H_{19}O_2)_3]$: C, 65.59; H, 9.43. Found: C, 64.89; H, 9.42.

Properties

The compound $[Mn(tod)_3]$ is an air-stable, crystalline solid that melts at 88–90°. The infared spectrum of this compound shows absorbances at 3000, 1550–1600, 1500, 1350–1410, 1220, 1150, 970, 900, 870 and 770 cm^{-1}. The mass spectrum exhibits a molecular ion at 604 a.m.u.

References

1. R. W. Moshier and R. E. Sievers, *Gas Chromatography of Metal Chelates*, Pergamon, Oxford, 1965.
2. R. L. Van Hemert, L. B. Spendlove, and R. E. Sievers, *J. Electrochem. Soc.*, **112,** 1123 (1965).

*The checkers report that, upon filtration, they obtained only some light-colored residue and a dark filtrate. Hence, the filtrate was stripped to near dryness on a Rotavap, and ~100 mL of hexane was subsequently added. The resultant solution was again filtered to remove the insoluble solids. The dark filtrate was mixed with ~300 mL of 95% EtOH in a beaker. The resultant solution was blown with a slow stream of air. When almost all the solvents were evaporated, a dark oil was obtained. Upon adding ~200 mL of H_2O and triturating, the crude product solidified and floated on water. The dark solid was collected and dried under vacuum. The crude yield was 14.7 g.

3. H. Koshimura, *J. Inorg. Nucl. Chem.*, **38,** 1705 (1976).
4. Y. A. Zolotov, *Extraction of Chelate Compounds,* translated by J. Schmorak, Humphrey Science Publ., Ann Arbor, MI, 1970.
5. Preliminary Report: LIX 54, A New Reagent for Metal Extractions from Ammoniacal Solutions, available from Henkel Corp.
6. R. L. Tischer, K. L. Eisentraut, K. Scheller, R. E. Sievers, R. C. Bausman, and P. R. Blum, Rep. No. ARL-74-0170, 1974.
7. T. J. Wenzel and R. E. Sievers, *U.S. Patent* No. 4,251,233 (1981).
8. R. Hauser, F. W. Swamer, and J. T. Adams, *Org. Reactions,* **8,** 59 (1954).
9. A. Leo, C. Hansch, and D. Elkins, *Chem. Rev.*, **71 (6),** 525 (1971).

29. LANTHANOID CROWN ETHER COMPLEXES

Submitted by D. WESSNER* and J.-C. G. BÜNZLI†
Checked by F. DUNBAR† and G. R. CHOPPIN†

$$LnX_3 + L \longrightarrow LnX_3{\cdot}L$$

X = NO_3; L = **1,2** (see below); Ln = La − Lu

L = **3;** Ln = La − Nd, Eu, Dy − Lu

$$4\ Ln(NO_3)_3 + 3L \longrightarrow [Ln(NO_3)_3]_4{\cdot}(L)_3$$

L = **3;** Ln = Nd − Gd, Dy

$$4\ Ln(NO_3)_3{\cdot}L \longrightarrow [Ln(NO_3)_3]_4{\cdot}(L)_3 + L$$

L = **2;** Ln = Gd − Lu

L = **3;** Ln = La − Pr, Tb, Ho − Lu

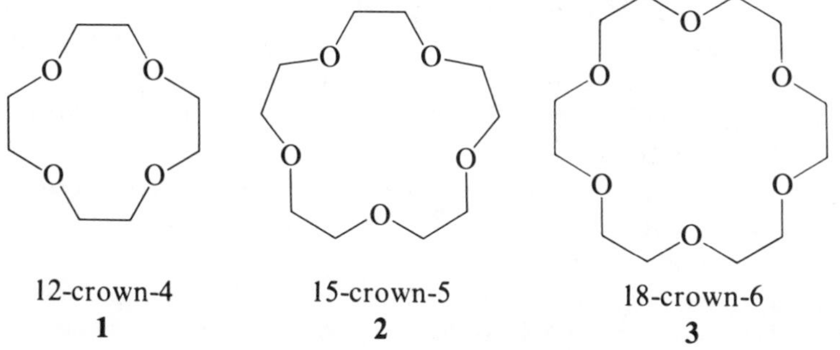

12-crown-4 **1**   15-crown-5 **2**   18-crown-6 **3**

*University of Lausanne, Institut de Chimie Minérale et Analytique, Place du Château 3, CH-1005 Lausanne, Switzerland.

†Department of Chemistry, The Florida State University, Tallahassee, FL 32306.

Interest in electrically neutral ion-complexing agents, such as macrocyclic polyethers (crown ethers), has increased continually since the original work of Pedersen.[1] Indeed, their selective metal binding properties[2] make them useful ligands for the study of the coordinative properties of metallic ions, for catalyzing synthetic processes, for analytical purposes[3] and for model studies of bioinorganic processes, such as the selective transport of cations across biological membranes.[4]

Recently, attention has been focused on lanthanoid/crown ether complexes,[5] since they may be used for lanthanoid ion separation, for stabilizing unusual oxidation states, and for studying high coordination numbers, Moreover, lanthanoid(III) ions are increasingly used as spectroscopic probes in systems of biological interest.[6]

The complexes between lanthanoid salts and crown ethers are usually isolated from nonaqueous solutions, the Ln(III)/polyether interaction being very small in water due to unfavorable energetics in removing water molecules from the inner coordination sphere of the metal ion. We report here the synthesis and the properties of complexes with 12-crown-4, 15-crown-5, and 18-crown-6 ethers,* having different metal/crown ratios, namely 1:1, 1:2, and 4:3. The single-crystal X-ray structures of four complexes have been solved. In the 1:1 complexes $Eu(NO_3)_3\cdot$(12-crown-4) and $Eu(NO_3)_3\cdot$(15-crown-5), which crystallize in a chiral space group, and $Nd(NO_3)_3\cdot$(18-crown-6), the metal ion displays coordination numbers of 10, 11, and 12, respectively. The first two complexes have similar structures: the polyether sits to one side of the Eu(III) ions and the three bidentate nitrate groups are coordinated on the opposite side. The structure of $[Nd(NO_3)_3]_4\cdot$(18-crown-6)$_3$ revealed that this complex has to be formulated as $[Nd(NO_3)_2\cdot$(18-crown-6)$]_3[Nd(NO_3)_3]_6$.

General

Except for Ln = Ce, the lanthanoid salts are prepared from the corresponding oxides (Glucydur, 99.99%) and concentrated acids (Merck). The solutions are filtered, evaporated to dryness, and the salts are dried for three days in a desiccator (KOH) and then for three days under vacuum ($60°/2 \times 10^{-2}$ torr). The salts contain one to five molecules of H_2O per formula weight.

The solvent (CH_3CN, Fluka) and the crown ethers (Fluka, purum) are used without further purification.

■ **Caution.** *12-crown-4 ether is believed to have some carcinogenic potency and operations with this ether and with liquid polyethers should be done with extreme care in a hood.*

The reactions are routinely performed under nitrogen atmosphere, eventhough many systems are stable to air and moisture. All the operations with the hygroscopic complexes are performed in a glove-box under an inert atmosphere

*12-crown-4 = 1,4,7,10-tetraoxacyclododecane; 15-crown-5 = 1,4,7,10,13-pentaoxacyclopentadecane; 18-crown-6 = 1,4,7,10,13,16-hexaoxacyclooctadecane.

(N_2, < 10 ppm H_2O). All the complexes are microcrystalline powders with the characteristic color of the lanthanoid ions. Complexometric analyses are performed with edta (Titriplex III, Merck), using xylenol orange and urotropine.

A. $Ln(NO_3)_3\cdot$(12-crown-4), Ln = La – Lu[7]

Procedure

A solution of 5 mmoles of 12-crown-4 in 50 mL of CH_3CN is added dropwise to a solution of 5 mmoles of $Ln(NO_3)_3\cdot xH_2O$ in 50 mL of CH_3CN. The resulting mixture is stirred at 60° for 24 hours. The polycrystalline complexes are filtered off after cooling, washed with CH_2Cl_2, and dried for 24 hours in a desiccator (P_4O_{10}). Anhydrous complexes are obtained for Ln = Nd-Lu. For Ln = La-Pr, monohydrated complexes are isolated, which can be further dried under high vacuum (48 h/10^{-5} torr). Yields: 70–85% for Ln = La-Er and 35–60% for Ln = Tm-Lu.

Properties

All the complexes have a 1:1 Ln(III)/crown stoichiometry. They are stable to air and moisture, except for Ln = La-Pr. They are thermally stable up to 280°, and then decompose completely, mainly into lanthanoid oxonitrates. The complexes have identical X-ray powder diagrams (CuK_α line) for Ln = Nd-Lu, and similar IR spectra for Ln = La-Lu. The main IR absorption frequencies in Nujol mulls are ~1495 (b), ~1280 (b), 1069, 1032, 1027, 1019, 930, 866, 819, and 742 cm^{-1} for Ln = Sm. The Raman spectra (Ar line, λ = 514.5 nm) of the finely powdered complex of samarium shows absorptions at 1032, 1014, 858, and 738 cm^{-1}. The corrected magnetic moments at 21° (Faraday's method) of the lighter Ln(III) ions in the 12-crown-4 complexes are 2.41, 3.42, 3.37, 1.66, 3.31, and 7.97 BM (Bohr Magnetons) for Ln = Ce, Pr, Nd, Sm, Eu, Gd, respectively. The complexes are completely dissociated in water and methanol, and they are slightly soluble in acetonitrile, in which conductimetric measurements (25°, 10^{-3} *M* solutions) indicate that all the nitrate groups are coordinated to the Ln(III) ions: Λ_M = 14 ohm^{-1} cm^2 $mole^{-1}$

B. $Ln(NO_3)_3\cdot$(15-crown-5), Ln = La – Lu[8,9]

Procedure

The same procedure is used as that described in Section A. For Ln = La-Eu, the compounds are anhydrous and not hygroscopic. For the heavier lanthanoids, the solutions are concentrated to 20–25 mL (Ln = Gd, Tb, Ho) or 50 mL

(Ln = Dy, Er-Lu). The complexes are dried over P_4O_{10}. They still contain three to four molecules of H_2O per formula weight. Completely anhydrous complexes can be obtained for Ln = Gd-Er after drying for a few days under high vacuum at −5° (until the pressure remains constant, ~10^{-5} torr). For Ln = Tm-Lu, part of the ligand is lost under high vacuum, and unsolvated complexes cannot be obtained. Yields: 60–85% for Ln = La-Eu, Tm-Lu; 20–40% for Ln = Gd-Er.

Caution. *For Ln = Gd-Lu, heating the complexes or drying under high vacuum for too long a time can partially decompose them into the 4:3 compounds.*

Properties

For Ln = La-Eu, the complexes are stable to air and moisture. They are thermally stable up to 260° (La), 240° (Pr), and 170° (Eu), and then decompose completely, mainly into oxonitrates. For the very hygroscopic complexes of the heavier lanthanoids, the thermogravimetric curves (Thermoanalyser Mettler, Ar flow, 2°/min, 10-mg samples) show, after loss of water, the formation of anhydrous 1:1 complexes of limited thermal stability. Further heating causes their quantitative transformation into the 4:3 complexes, which decompose above 250–270 °. The thermogravimetric data are reported in Table I.

X-ray powder photographs of some complexes indicate at least three types of structures within the series: Ln = La-Nd, Ln = Sm, Eu, Ln = Ho. The IR spectra are similar for Ln = La-Nd; the main absorptions are at 1500, 1360, 1310, 1075, 1035, 968, 960, 881, and 741 cm^{-1}; Raman bands: 1500, 1470, 1460, 1270, 1034, 1031, 870, and 732 cm^{-1}. The IR spectra for Ln = Sm-Er are similar; the main IR absorptions (Ho) are at 1500, 1354, 1320, 1275, 1250, 1130, 1092, 1060, 1025, 950, 878, 816, and 748 cm^{-1}. The corrected magnetic moments (21°) of the Ln(III) ions are 2.36, 3.33, 3.34, 1.58, 3.31, and 7.88

TABLE I Thermogravimetric Data for $Ln(NO_3)_3 \cdot (15\text{-crown-5}) \cdot nH_2O$ Complexes

Ln	n	1:1 stab. range	4:3 stab. range
La	0	≤260°	
Pr	0	≤240°	
Eu	0	≤170°	
Gd	4.2	100–160°	195–250°
Tb	2.0	100–110°	155–260°
Dy	4.6	(90°)	150–260°
Ho	5.1	(90°)	150–270°
Er	3.4	105–110°	170–275°
Yb	3.9	(105°)	165–270°

BM for Ln = Ce, Pr, Nd, Sm, Eu, and Gd, respectively. The $Ln(NO_3)_3 \cdot (15$-5) complexes are slightly soluble in CH_3CN; the molar conductivity (25°, 10^{-3} *M* solution) of the praseodymium complex in CH_3CN is 9 ohm^{-1} cm^2 $mole^{-1}$.

C. $[Ln(NO_3)_3]_4 \cdot (15\text{-crown-}5)_3$, Ln = Gd – Lu[9]

$$4\ Ln(NO_3)_3 \cdot (15\text{-}5) \cdot xH_2O \xrightarrow{\Delta} [Ln(NO_3)_3]_4 \cdot (15\text{-}5)_3 + 15\text{-}5 + 4x\ H_2O$$

Procedure

The 4:3 complexes are obtained by quantitative thermal decomposition under high vacuum of the corresponding 1:1 complexes. The experimental conditions are reported in Table II.

Properties

The $[Ln(NO_3)_3]_4 \cdot (15\text{-crown-}5)_3$ complexes, Ln = Gd – Lu, are very hygroscopic. The X-ray powder photographs of the 4:3 and 1:1 complexes of holmium are identical. The IR spectra of the series are similar, with the main absorptions being at 1320, 1278, 1250, 1100, 1062, 1025, 952, 880, 819, and 751 cm^{-1}.

D. $Ln(NO_3)_3 \cdot (18\text{-crown-}6)$, Ln = La – Nd, Eu, Tb – Lu[8,9]

Procedure

1. Ln = La – Pr, Er – Lu

The same procedure as described in Section A is used. For Ln = La-Pr, the compounds are anhydrous and not hygroscopic. For Ln = Er-Lu, complexes with ~three molecules of H_2O per formula weight are obtained, after drying for 24 hours over P_4O_{10}. These complexes are dried at room temperature under high

TABLE II Conditions for Quantitative Thermal Transformation of $Ln(NO_3)_3 \cdot (15\text{-crown-}5)$ into $[Ln(NO_3)_3]_4 \cdot (15\text{-crown-}5)_3$ at 10^{-5} torr

Ln	t(hr)	T	Ln	t(hr)	T
Gd	24	160°	Er	96	25°
Tb	5	120°	Tm	24	40°
Dy	120	25°	Yb	96	25°
Ho	3	80°	Lu	72	40°

vacuum for 2–4 days until the pressure remains constant ($\sim 10^{-5}$ torr). Yields: 40% for Ln = Er, 50–65% for Ln = La-Pr, Yb, 75–90% for Ln = Tm, Lu.

2. Ln = Nd

The filtrate of the neodymium solution (see Section E) is concentrated to 75 mL, and it yields a second crystalline fraction corresponding to the 1:1 complex, which is dried over P_4O_{10}. This compound is anhydrous, but not hygroscopic. Yield: 30%.

3. Ln = Tb, Ho

Exactly 5 mmoles of $Ln(NO_3)_3 \cdot xH_2O$ in 20–40 mL of acetonitrile are added dropwise at 0° to 5 mmoles of 18-crown-6 in 75 mL of CH_3CN. The resulting solution is stirred for 24 hours at 0° and is filtered. The complexes are dried for a few days over P_4O_{10} and for three to four days at room temperature under high vacuum, until the pressure remains constant ($\sim 10^{-5}$ torr). These compounds are hygroscopic. Yield: 35%.

4. Ln = Eu, Dy

Exactly 5 mmoles of $Ln(NO_3)_3 \cdot xH_2O$ in 80 mL of CH_3CN is added dropwise at 0° to 5 mmoles of 18-crown-6 in 75 mL of CH_3CN. The resulting solution is stirred for 24 hours at 0°, filtered, stirred for three days at −20°, and filtered. The complexes are dried for a few days over P_4O_{10} and for 3–4 days at room temperature under high vacuum until the pressure remains constant ($\sim 10^{-5}$ torr). These compounds are hygroscopic. Yield: 30%. For Ln = Sm, Gd, the same procedure leads to 4:3 complexes.
Caution. *For Ln = Eu, Tb − Lu, heating the complexes or drying under high vacuum for too long a time can partially decompose them into 4:3 compounds.*

Properties

The complexes with Ln = La-Nd are stable to air and moisture, but the compounds of the heavier lanthanoids are hygroscopic. The thermogravimetric data (Table III) show that hydrated compounds lose water to give anhydrous 1:1 complexes. The 1:1 complexes (Ln = La-Nd, Eu, Tb-Lu) are then quantitatively transformed into 4:3 complexes, which decompose completely at higher temperature, mainly into oxonitrates.

The heats of the endothermic 1:1 → 4:3 transformation measured by dif-

TABLE III Thermogravimetric Data for $Ln(NO_3)_3$·(18-crown-6)·nH_2O Complexes

Ln	n	1:1 stab. range	4:3 stab. range
La	0	≤ 210°	230–305°
Pr	0	≤ 130°	220–290°
Nd	0	≤ 110°	180–250°
Eu	0.4	50–110°	—[a]
Tb	3.5	100–110°	—[a]
Er	3.1	105–135°	200–270°
Yb	3.5	105–120°	195–235°

[a]Undefined plateau

ferential scanning calorimetry (0.5°/min, samples of 20–40 mg in sealed ampulla) are 57, 47.2, 43.4, and 37.6 kJ $mole^{-1}$ for Ln = La, Ce, Pr and Nd, respectively. For Ln = La, Pr, and Nd, the X-ray powder photographs are identical; they are different from the photograph of the holmium complex.

The main IR absorptions for Ln = La-Nd are 1490, 1360, 1312, 1085, 1048, 967, 949, 880 (medium intensity), 847, 821, and 745 cm^{-1}; Raman bands: 1475, 1450, 1440, 1277, 1245, 1045, and 873 cm^{-1}. The main IR absorptions for Ln = Eu, Tb-Lu are 1515, 1334, 1280, 1110, 1070, and 967 cm^{-1}; Raman bands: 1465, 1045, 1025, and 878 cm^{-1}. These complexes do not show an IR absorption at 880 cm^{-1}. The corrected magnetic moments (21°) of the Ln(III) ions are 2.38, 3.33, and 3.37 BM for Ln = Ce, Pr and Nd, respectively.

The 1:1 complexes of $Ln(NO_3)_3$ and 18-crown-6 are completely dissociated in water, and they are slightly soluble in CH_3CN (~0.02 *M*), in which they are essentially non-conductors (Λ = 28 and 57 ohm^{-1} cm^2 mol^{-1} at 25° for 10^{-3} *M* solutions of Pr and Nd complexes, respectively). The ^{1}H-nmr in CD_3CN show the presence, at room temperature, of both free (δ = ~3.6 ppm from TMS) and complexed 18-crown-6 (δ = 3.85, −2.95, −8.05, −0.53, 8.57, and 35.95 ppm for Ln = La, Ce, Pr, Nd, Eu and Yb, respectively). The formation constants in CD_3CN, measured by ^{1}H-nmr, decrease when the atomic number increases (log K_f ± 0.2 = 4.4, 4.5, 3.7, 3.5, 2.6, and 2.3 for Ln = La, Ce, Pr, Nd, Eu, and Yb, respectively).

E. $[Ln(NO_3)_3]_4$·(18-crown-6)$_3$, Ln = La − Lu[8,9]

The 4:3 complexes of the 18-crown-6 ether are obtained either by direct crystallization from CH_3CN solutions (Ln = Nd-Gd, Dy) or by quantitative thermal decomposition of the 1:1 complexes (Ln = La-Pr, Tb, Ho-Lu).

Procedure

1. Ln = Nd-Gd, Dy

A solution of 5 mmoles of 18-crown-6 in 50 mL of CH_3CN is added dropwise to a solution of 5 mmoles $Ln(NO_3)_3 \cdot xH_2O$ in 100 mL CH_3CN. The resulting mixture is stirred at 60° for 24 hours. For Ln = Gd, Dy, the solution is evaporated to 25 mL. The polycrystalline complexes are filtered off after cooling, washed with CH_2Cl_2, and dried for 24 hours over P_4O_{10}. For Ln = Nd-Eu, the complexes are anhydrous and not hygroscopic. The complexes of Gd and Dy are dried for three days (10^{-5} torr/25°). Yields: 15% for Ln = Nd, 30–45% for Ln = Sm-Gd, Dy.

2. Ln = La-Pr, Tb, Ho-Lu

The experimental conditions for the quantitative thermal decompositions are reported in Table IV. The complexes with Ln = Tb, Ho-Lu, are hygroscopic.

Properties

The complexes with Ln = Gd-Lu are very hygroscopic. The X-ray powder photographs are identical for Ln = Ce, Nd, Eu, Ho. The main IR absorptions are at 1510, 1332, 1278, 1067, and 966 cm^{-1}; Raman bands: 1520, 1462, 1045, 1028, and 880 cm^{-1}. The corrected magnetic moments (21°) are 2.35, 3.35, 3.32, 1.59, 3.30, and 7.89 BM for Ln = Ce, Pr, Nd, Sm, Eu, and Gd, respectively. The molar conductivities in CH_3CN, (25°, 10^{-3} M solutions) are 44 and 66 ohm^{-1} cm^2 $mole^{-1}$ for the Pr and Nd complexes, respectively.

TABLE IV Conditions for Quantitative Thermal Transformation of $Ln(NO_3)_3 \cdot$(18-crown-6) into $[Ln(NO_3)_3]_4 \cdot$(18-crown-6)$_3$

Ln	Time (hr)	Pressure (torr)	Temperature
La	24	2×10^{-2}	225°
Ce	22	2×10^{-2}	190°
Pr	16	720	170°
Tb	48	5×10^{-5}	60°
Ho	24	2×10^{-2}	110°
Er	16	2×10^{-2}	180°
Tm	96	5×10^{-5}	40°
Yb	96	2×10^{-2}	70°
Lu	48	5×10^{-5}	40°

References

1. C. J. Pedersen, *J. Am. Chem. Soc.*, **89,** 7017 (1967).
2. J. J. Christensen, D. J. Eatough, and R. M. Izatt, *Chem. Rev.*, **74,** 351 (1974).
3. I. M. Kolthoff, *Anal. Chem.*, **51,** 1R (1979).
4. J. J. Grimaldi and J.-M. Lehn, *J. Am. Chem. Soc.*, **101,** 1334 (1979).
5. R. B. King and P. R. Heckley, *J. Am. Chem. Soc.*, **96,** 3118 (1974); J.-C. G. Bunzzi and D. Wessner, *Coord. Chem. Revs.*, **60,** 191 (1984).
6. J. Reuben, in *Handbook on the Physics and Chemistry of Rare Earths,* K. A. Gschneidner and L. Eyring (ed.), North-Holland Publ. Co., Amsterdam, 1979, Chapter 39.
7. J.-C. G. Bünzli and D. Wessner, *Inorg. Chim. Acta,* **44,** L55 (1980).
8. J.-C. G. Bünzli and D. Wessner, *Helv. Chim. Acta,* **61,** 1454 (1978).
9. J.-C. G. Bünzli and D. Wessner, *Helv. Chim. Acta,* **64,** 582 (1981).

30. *POLY*-BIS(THIOCYANATO-*N*)BIS-μ-(1*H*-1,2,4-TRIAZOLE-N^2,N^4)METAL(II) COMPLEXES WITH MANGANESE, IRON, COBALT, NICKEL, COPPER AND ZINC, AND BIS[TRIS(THIOCYANATO-*N*)TRIS(μ-4*H*-1,2,4-TRIAZOLE-N^1)NICKEL(II)-N^2,$N^{2'}$,$N^{2''}$]NICKEL(II)

Submitted by J. G. HAASNOOT,* D. W. ENGELFRIET,* and J. REEDIJK*
Checked by JAMES T. CRONIN,† J. A. McLAREN,† K. J. STAUFFER,† J. A. TEAGLE,† and J. L. BURMEISTER†

The synthesis of ''low-dimensional solids'' has become more important than ever before, both to chemists and physicists. In the catenation of metal ions, the catenating blocks are usually polynucleating ligands. The 1,2,4-triazole ligand acts in most of its metal complexes as a 1,2-bicoordinating ligand. In this type of coordination, linear polynuclear complexes are formed.

The thiocyanates of first-row transition elements, however, give compounds of overall composition of $M(NCS)_2(trz)_2$ in which the triazole molecules are 2,4-bicoordinating. The obtained geometry is a two-dimensional array of metal ions linked by triazole bridges. The crystal structures of these complexes have been investigated by Engelfriet and co-workers.[1] The complexes with paramagnetic metal ions afford magnetically interesting systems.[2] For some of these compounds, monocrystals up to several mm long can easily be prepared.

With nickel, a linear trinuclear α-isomer $[Ni(NCS)_2(trz)_2]$ can be prepared,

*Gorlaeus Laboratories, Dept. of Chemistry, State University Leiden, P.O. Box 9502, 2300 RA Leiden, The Netherlands.

†Department of Chemistry, University of Delaware, Newark, DE 19711.

and, by slight variation of the reaction conditions, the two-dimensional β-isomer can be produced.[3] Because of the symmetry of the triazole ring in its 1H and 4H tautomeric forms, the difference between the isomers can be seen easily from the infrared spectra.[4] A third isomer (γ-isomer) of the cadmium compound was recently found by this group, in which the triazole molecules coordinate monodentately through N^4 and the cadmium ions are bridged, forming linear chains by N,S-coordinating thiocyanate ions.[5]

$$\text{n M(NCS)}_2 + \text{2n trz} \longrightarrow [\text{M(NCS)}_2(\text{trz})_2]_\text{n}$$

A. PROCEDURE FOR $[Mn(NCS)_2(trz)_2]_n$

Manganese(II) thiocyanate (0.05 mole, 8.55 g) is dissolved in 100 mL of water. Then 1,2,4-triazole (0.1 mole 6.91 g), dissolved in 50 mL of water, is added. A few drops of concentrated nitric acid are added and the solution is heated to boiling, cooled immediately to room temperature and stored for several days in a desiccator above phosphorus pentoxide. The yield of the product, recovered by filtration, washed with water and dried under vacuum is ~12 g or about 75%. The crystals have the shape of flat rhombi, almost squares, with a thickness of about one-fifth of the largest diameter. Crystal diameters may vary from 1–10 mm depending upon the preparative technique. Crystallization of this compound occurs more slowly than for the compounds containing other metals.

B. PROCEDURE FOR $[Fe(NCS)_2(trz)_2]_n$

The iron compound can be prepared from any soluble iron(II) salt. It is most convenient to start with hydrated iron(II) perchlorate. Hexaaquairon(II) perchlorate (0.05 mole, 18.14 g) is dissolved in about 100 mL of degassed water. Ammonium thiocyanate (0.1 mole 7.61 g) dissolved in about 50 mL of degassed water is added; the color of the solution turns red as a result of small amounts of iron(III) present. This solution is saturated with sulfur dioxide (to reduce any Fe(III) present) and a few drops of nitric acid are added. Next, 1,2,4-triazole (0.1 mole, 6.91 g), dissolved in 50 mL water, is added. At this stage the color of the solution may still be (dark) brown.

By heating the solution to boiling, the color changes to faint green (or colorless), indicating the reduction of iron(III) to iron(II); at the same time, the excess sulfur dioxide is removed by evaporation. Prolonged boiling renders the solution vulnerable to atmospheric oxygen. When required, additional sulfur dioxide is passed into the solution, followed by heating, until decoloration occurs. Upon cooling to room temperature, a white precipitate is produced. This is filtered and washed with oxygen-free water containing a small amount of sulfur dioxide. The product is dried under vacuum. The yield varies, depending upon the amounts of water and acid, but usually is about 10 g (~65%). Large crystals, up to 3 mm long, can be grown by evaporation of dilute solutions under a nitrogen atmosphere in a desiccator above phosphorus pentoxide.

C. PROCEDURE FOR $[Co(NCS)_2(trz)_2]_n$

The cobalt compound is prepared in a manner analogous to that of the manganese compound, using about twice as much water and acid to dissolve the initially produced gelatinous precipitate. The yield is ~12 g (75%). Crystals up to 5 mm long can be grown from dilute solutions by slow evaporation in a desiccator above phosphorus pentoxide.

D. PROCEDURE FOR β-$[Ni(NCS)_2(trz)_2]_n$

Ammonium thiocyanate (0.1 mole, 7.61 g), dissolved in 50 mL water, is added to nickel(II) nitrate, $Ni(NO_3)_2 \cdot 6H_2O$ (0.05 mole, 14.54 g), in 100 mL water. The solution turns dark green and 2 mL of concentrated nitric acid is added. Then, 1,2,4-triazole (0.1 mole, 6.91 g) in 50 mL water, is added. The dark blue solution is left standing for several days, whereupon purple-blue crystals of β-$[Ni(NCS)_2(trz)_2]$ are formed. The crystals tend to be rather small, up to 0.1 mm long. The product is recovered by filtration, washed with water, and dried under vacuum. The yield is 13.5 g (about 86%).

E. PROCEDURE FOR $[Cu(NCS)_2(trz)_2]_n$

$[Cu(NCS)_2(trz)_2]$ cannot be prepared from commercial $Cu(NCS)_2$ because this substance is rather insoluble in most solvents and is usually very impure. A useful procedure is to start with freshly precipitated $Cu(NCS)_2$, for instance from copper(II) nitrate and ammonium thiocyanate in water. Ammonium thiocyanate (0.2 mole, 15.22 g) in 50 mL of water is added to copper (II) nitrate trihydrate (0.1 mole, 24.16 g) in 100 mL of water. Copper(II) thiocyanate is precipitated immediately. After filtration and washing with water, it is dried under vacuum. The precipitate should be pitch-black; any gray color is a sign of copper(I) thiocyanate.

Thus prepared, copper(II) thiocyanate (0.05 mole, 8.99 g) is suspended in 100 mL of warm acetone. Then, 1,2,4-triazole (0.1 mole, 6.91 g) in 100 mL of warm acetone is added, and, after some stirring, the mixture is filtered and the residue is washed with warm acetone. The residue is a mixture of copper(II) thiocyanate, copper(I) thiocyanate, and thiocyanato-1,2,4-triazolatocopper(II). The combined filtrates are evaporated until almost dry, and, to this residue, 20mL of ethanol is added, whereupon a green precipitate is produced. After filtration, washing with ethanol, and drying under vacuum, $[Cu(NCS)_2(trz)_2]$ can be recovered as tiny blue-green crystals; yield ~12 g (75%). Crystals up to 0.5 mm long can be grown from acetone solution.

F. PROCEDURE FOR $[Zn(NCS)_2(trz)_2]_n$

The zinc complex is prepared in a manner analogous to that of the manganese complex. The yield is lower due to the greater solubility of the product. The crystal size depends on the amount of water and acid present in the solution, but may be up to 1 mm long.

$$3\ Ni^{++} + 6\ trz + 6\ NCS \longrightarrow \alpha\text{-}[Ni(NCS)_2(trz)_2]_3$$

G. PROCEDURE FOR α-$[Ni(NCS)_2(trz)_2]_3$

The 1,2,4-triazole ligand (0.1 mole, 6.91 g) in 50 mL water is added to nickel(II) nitrate (0.05 mole, 14.54 g of the hexahydrate) dissolved in 50 mL water, whereupon the solution turns dark blue. Then, ammonium thiocyanate (0.1 mole, 7.61 g) in 25 mL water is added. Any turbidity is removed by filtration. Upon standing for several days, purple crystals of α-$[Ni(NCS)_2(trz)_2]$ are formed. The crystal size depends on the concentration of reactants and the temperature, and may be up to 0.2 mm long. The product is recovered by filtration, washed with water and dried under vacuum. The yield is about 14 g (89%).

Characterization

All compounds can be characterized by their infrared spectra. The spectra show the layer compounds to be isomorphous and to differ distinctly from the spectrum of α-$[Ni(NCS)_2(trz)_2]$. The region 600–700 cm^{-1} is especially informative. One strong absorption at 630 cm^{-1} occurs for α-$[Ni(NCS)_2(trz)_2]$, while two strong absorptions occur for the layer compounds at 640 cm^{-1} and 670 cm^{-1}. Products with only one strong absorption at 670 cm^{-1} contain large amounts of triazolate ions, which are usually produced in neutral or basic solutions.[2]

References

1. D. W. Engelfriet, W. den Brinker, G. C. Verschoor, and S. Gorter, *Acta Cryst.*, **B35,** 2922 (1979).
2. D. W. Engelfriet, Thesis, Leiden (1980).
3. D. W. Engelfriet, J. G. Haasnoot, and W. L. Groeneveld, *Z. Naturforsch,* **32a,** 783 (1977).
4. J. G. Haasnoot and W. L. Groeneveld, *Z. Naturforsch,* **32b,** 553 (1977).
5. J. G. Haasnoot, G. C. M. de Keyzer, and G. C. Verschoor, *Acta Cryst.*, **C39,** 1207 (1983).

31. PREPARATION OF PENTAINDIUM TETRASULFIDE IN LIQUID TIN

$$5\ \mathrm{In} + 4\ \mathrm{S} \xrightarrow[\mathrm{Sn(l)}]{} \mathrm{In_5S_4(s)}$$

Submitted by T. WADSTEN*
Checked by A. WOLD and V. K. KRIEBLE†

In the indium-sulfur system the phases InS[1], In_6S_7[2], and three allotropic forms of In_2S_3[3,4,5] have been identified and their structures analyzed. All of these compounds have been prepared by conventional methods, that is, direct combination or transport reaction techniques. By using molten tin as a solvent, a new metal-rich combination was obtained. The material is electrically insulating and diamagnetic at room temperature.

Procedure

■ **Caution.** *Proper shielding should be used when heating sealed tubes.* Mixtures of indium, sulfur and tin, all of high purity, are sealed in evacuated transparent silica tubes. The In:S ratio should be close to 1:1, and excess tin, four times the weight of the reactants, is added. This is close to the solute-to-solvent ratio that was found to give the best crystalline yield in the preparation of gallium monochalcogenides.[6] The tubes are heated at a constant rate of 50°/hr up to temperatures of 650–1100°.

To assure complete dissolution of the elements, the reactants are held at the maximum temperature for a few hours before the temperature decrease is started.

*University of Stockholm, Department of Inorganic Chemistry, Arrhenius Laboratory, S-106 91 Stockholm, Sweden.

†Department of Chemistry, Brown University, Providence, RI 02912.

After having reached room temperature the ampoules are cut open and the reaction products further treated. The products obtained (of which three main types can be distinguished) are found to be highly dependent on the cooling program.

The first one is a rapid procedure, about 50°/hr down to 650°, followed by very slow cooling to room temperature, at a few degrees per hour. This method produces a homogeneous batch of large In_5S_4 crystals. The second type of cooling program is the converse of the first one, that is, very slow cooling down to 650°, and then rapid cooling (50°/hr). In this case the solid residue consists of black rods with the composition In_6S_7. The third procedure uses an intermediate cooling rate of about 25°/hr from the maximum temperature all the way down to room temperature, and yields a mixture of polycrystalline In_5S_4 and In_6S_7.

A high maximum temperature (~1100°) is found to give the best yield of the products. The lowest temperature that yields In_5S_4 is 650°. To isolate the reaction products from the tin matrix, the chunks of reacted material are treated with concentrated hydrochloric acid. To reduce the attack on the crystals, short-time treatments with acid are alternated with acetone washings.

Another technique, more time-consuming but chemically not so aggressive as the acid treatment, is to dissolve the tin in benzene containing 10% iodine. After the procedure, the solid product may be coated with a very thin amorphous brownish product of unknown composition. The coating material does not give any extra lines in the X-ray powder diffraction photographs, however.

The sulfur content of In_5S_4 was determined by gas chromatographic techniques devised by Kirsten.[7] The observed value is 18.5 ± 0.2% by weight, and the calculated value is 18.3% for In_5S_4. By X-ray fluorescence methods, no impurities, such as tin, are observed. The detection limit is estimated to be of the order of 100 ppm.

Properties

The In_5S_4 crystals have been found to form in different shapes depending on the thermal pre-history. The X-ray powder patterns of the reaction batches were identical, however, and all the crystals are red and transparent. The cubic unit cell dimension is a = 12.35 Å, and the cell contains eight formula units of In_5S_4. The structure is of a new type.[8] The calculated density is 4.95 g/L compared with the observed density of 4.87 g/L. The magnetic susceptibility was found to be -1.9×10^{-4} $cm^3 mole^{-1}$.

References

1. K. Schubert, E. Dörre, and E. Günzel, *Naturwiss.*, **41,** 448 (1954).
2. J. H. C. Hogg and W. J. Duffin, *Acta Cryst.*, **23,** 111 (1967).

3. H. Hahn and W. Klingler, *Z. Anorg. Chem.*, **260,** 97 (1949).
4. G. A. Stiegman, H. H. Sutherland, and J. Goodyear, *Acta Cryst.*, **19,** 967 (1965).
5. R. Diehl, C.-D. Carpentier, and R. Nitsche, *Acta Cryst.*, **B32,** 1257 (1976).
6. T. Wadsten, *Chemica Scripta,* **8,** 63 (1975).
7. W. J. Kirsten, *Z. Anal. Chem.*, **18,** 1 (1961).
8. T. Wadsten, L. Arnberg, and J.-E. Berg, *Acta Cryst.*, **B36,** 2220 (1980).

32. SALTS OF ALKYL[2-[1-[(2-AMINOETHYL)IMINO]ETHYL]PHENOLATO](1,2-ETHANEDIAMINE)COBALT(III) AND ALKYL[2-[1-[(3-AMINOPROPYL)IMINO]ETHYL]PHENOLATO](1,3-PROPANEDIAMINE)COBALT(III)

Submitted by I. YA. LEVITIN,* R. M. BODNAR,* and M. E. VOL'PIN*
Checked by G. N. SCHRAUZER,† N. H. LIU,† K. RUBIN,† and M. KARBASSI†

Until recently all known σ-organic derivatives of trivalent cobalt included either a planar tetradentate (or bis-bidentate) ligand or several specific monodentate or bidentate strong-field ligands, such as cyanide ion and 2,2′-bipyridine. The new type of organocobalt(III) complexes at issue, $[\mathbf{1}]^+X^-$, were reported[1] in 1981. They contain as ligands,

$$\{RCo^{III}[o\text{-}OC_6H_4C(Me){=}N(CH_2)_nNH_2][H_2N(CH_2)_nNH_2]\}^+X^- \quad [\mathbf{1}]^+X^-$$

$$RCo^{III}[o\text{-}OC_6H_4C(Me){=}N(CH_2)_nN{=}C(Me)C_6H_4(o\text{-})O] \quad \mathbf{2}$$

$$\mathbf{a} \text{ if } n = 2; \quad \mathbf{b} \text{ if } n = 3$$

apart from an alkyl group, a chelating diamine and a mixed tridentate ligand derived from the Schiff base constituted by *o*-hydroxyacetophenone and the same diamine in a 1:1 ratio. Evidence for the nature of the ligands in complexes of the new type is based on studies of IR and ^{1}H nmr spectra, as well as on analyses of degradation products.[1,2] Furthermore, the coordination geometry of the complex cations derived from 1,2-ethanediamine, $[\mathbf{1a}]^+$, that is, those with meri-

*Institute of Organo-Element Compounds, Academy of Sciences of the U.S.S.R., Moscow B-334, Union of Soviet Socialist Republics.
†Department of Chemistry, University of California, San Diego, La Jolla, CA 92093.

dional attachment of the tridentate ligand, is known from an X-ray single-crystal diffraction study.[1]

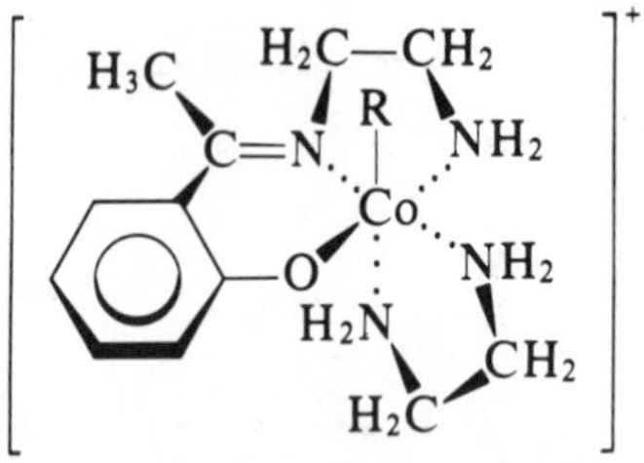

$[\mathbf{1a}]^+$

To synthesize the ''unsymmetrical'' cationic organometal complexes, [**1**]X, a single-step template technique is employed. The mainstream process can be represented as follows:

$$\text{(a)}\ \mathrm{Co^{II}Y_2 + 2\ HOC_6H_4COMe + 2\ H_2N(CH_2)_nNH_2 + 2\ NaOH \longrightarrow Co^{II}[OC_6H_4C(Me){=}N(CH_2)_nNH_2]_2 + 2\ NaY + 4\ H_2O}$$

$$\text{(b)}\ \mathrm{8\ Co^{II}[OC_6H_4C(Me){=}N(CH_2)_nNH_2]_2 + NaBH_4 + 8\ NaOH \xrightarrow{Pd} 8\ Na\{Co^{I}[OC_6H_4C(Me){=}N(CH_2)_nNH_2]_2\} + NaBO_2 + 6H_2O}$$

$$\text{(c)}\ \mathrm{Na\{Co^{I}[OC_6H_4C(Me){=}N(CH_2)_nNH_2]_2\} + RX \longrightarrow RCo^{III}[OC_6H_4C(Me){=}N(CH_2)_nNH_2]_2 + NaX}$$

$$\text{(d)}\ \mathrm{RCo^{III}[OC_6H_4C(Me){=}N(CH_2)_nNH_2]_2 + H_2O + NaX \longrightarrow \{RCo^{III}[OC_6H_4C(Me){=}N(CH_2)_nNH_2][H_2N(CH_2)_nNH_2]\}X + NaOC_6H_4COMe,^*}$$

where Y = Cl or OAc, n = 2 or 3, X = Br or I,† and R is primary or secondary.

*The ketophenolate undergoes further transformation yielding a resol-type oligomer:

$$\text{(e)}\ \mathrm{4\ NaOC_6H_4COMe + NaBH_4 + 2\ H_2O \longrightarrow 4\ NaOC_6H_4CHOHMe + NaBO_2}$$

$$\text{(f)}\ \mathrm{m\ NaOC_6H_4CHOHMe \longrightarrow H(NaOC_6H_3CHMe)_mOH + (m-1)\ H_2O}$$

†In addition, the perchlorate salts were obtained by way of anion exchange:

$$\mathrm{\{Co^{III}R[OC_6H_4C(Me){=}N(CH_2)_nNH_2][H_2N(CH_2)_nNH_2]\}X + NaClO_4 \longrightarrow \{Co^{III}R[OC_6H_4C(Me){=}N(CH_2)_nNH_2][H_2N(CH_2)_nNH_2]\}ClO_4 + NaX}$$

alkyl or cyclohexyl. The reactions (a) to (f) can be summarized in a single equation:

$$8\ Co^{II}Y_2 + 16\ HOC_6H_4COMe + 16\ H_2N(CH_2)_nNH_2 + 8\ RX + 3\ NaBH_4 + 24\ NaOH \longrightarrow$$
$$8\ \{RCo^{III}[OC_6H_4C(Me){=}N(CH_2)_nNH_2][H_2N(CH_2)_nNH_2]\}X + 8/m\ H(NaOC_6H_3CHMe)_mOH + 16\ NaY + 3\ NaBO_2 + (34 - 8/m)\ H_2O$$

"Symmetrical" neutral organometal complexes of the well-known type **2,** including the tetradentate Schiff base ligand of 2:1 stoichiometry,* are formed concurrently, but in minor amounts. It is also noteworthy that (a) the only organocobalt products isolated after similar syntheses starting from certain hydroxycarbonyl compounds other than *o*-hydroxyacetophenone (e.g., 2,4-pentanedione and salicylaldehyde) are of the "symmetrical" type,[4] and (b) an earlier attempt to prepare σ-organocobalt(III) complexes with chelating 1,2-ethanediamine failed.[5]

Finally, the rather unusual reactivity of the complexes at issue under conditions of protonolysis[1,2] should be mentioned.

A. ALKYL-[2-[1-[(2-AMINOETHYL)IMINO]ETHYL]-PHENOLATO](1,2-ETHANEDIAMINE)COBALT(III) BROMIDE (The Series of [1a]Br Complexes)

Procedures

■ **Caution.** *Since the preparations involve certain extremely hazardous, volatile substances, including o-hydroxyacetophenone, methanol, and 1,2-ethanediamine as starting materials, hydrogen is evolved during the course of the syntheses, and the compounds are especially malodorous, use of a ventilated hood is necessary.*

1. The Ethylcobalt Complex (R = Et)

This synthesis is carried out under anaerobic (pure nitrogen) conditions and with constant stirring. All protracted operations with the organocobalt product as well as with its solutions are performed in the dark.

The synthesis is conducted in a 250-mL three-necked flask fitted with an effective stirrer, a dropping funnel, and a T-joint connected to a nitrogen/vacuum line and a bubbler.

*Some of these complexes have been prepared by Costa and co-workers[3] by a different technique.

The flask is charged with 100 mL of deaerated methanol. Then, *o*-hydroxyacetophenone (3.4 mL, 3.8 g, 28 mmoles), 1,2-ethanediamine (2.4 mL of its 70% aqueous solution,* 1.70 g of the pure reagent, 28 mmoles), and cobalt(II) chloride hexahydrate† (3.3 g, 14 mmoles) are introduced with stirring. After dissolution of the salt, the flask is placed into a cold-water bath.* Then, 4.5 mL of a 50% aqueous solution of sodium hydroxide is added. The original wine-red solution turns into a carrot-colored suspension. After five minutes, about one third of a solution of sodium tetrahydroborate(-1) (*in toto* 0.75 g, 20 mmoles) in 5 mL of 5% aqueous sodium hydroxide (85 mmoles of pure NaOH is the total amount required for the synthesis) is added, followed by 0.25 mL of a 2% solution of palladium(II) chloride in 1 *M* aqueous potassium chloride. The suspension gradually darkens to some extent. After 15 minutes, the water bath is removed and the flask is shielded from light. Next, ethyl bromide (5 mL, 6.8 g, 65 mmoles) is introduced. The remaining solution of sodium tetrahydroborate(1-) is added in several portions over a period of five hours. Stirring is continued for three hours.

The brown suspension resulting from the synthesis is filtered through a fine, porous, glass filter, and the solid residue is extracted with methanol until the solvent remains colorless. The liquid is concentrated under vacuum to a volume of 35 mL, and is then diluted with 20 mL of water; a saturated aqueous solution of sodium bromide (10 mL) is also added. The resulting solution is concentrated to a volume of 25 mL in the same manner. The resulting precipitate is collected by filtration and washed with cold water, a small amount of acetone, and diethyl ether. Then, it is extracted with dichloromethane until the solvent‡ is no longer green. The residue is dried by suction, yielding 2.36 g (5.8 mmoles, 42%) of an orange-to-red crystalline material. The product is recrystallized from methanol. *Anal.* Calcd. for $C_{14}H_{26}ON_4BrCo$: C, 41.50; H, 6.47; N, 13.83; Br, 19.72; Co, 14.54. Found: C, 41.53; H, 6.31; N, 14.06; Br, 19.74; Co, 13.89.

2. Other Complexes of the Series ([**1a**]Br)

Related bromide salts with other primary and secondary alkyl ligands (R) can be obtained similarly and in comparable yields. Thus, the complexes with R = Me, Bu, *i*-Pr and cyclohexyl are prepared starting from the corresponding alkyl bromides. In the case of R = Me, gaseous methyl bromide§ is passed through

*The concentration of the solution is not critical; the anhydrous base or the hydrochloride can be used as well. In the latter case, an equivalent amount of sodium hydroxide must also be added (with cooling). In this procedure, the bath temperature should be kept below 40°.

†The acetate, $Co(OAc)_2 \cdot 4H_2O$, can be used instead.

‡The dichloromethane liquor contains the corresponding "symmetrical" complex **2**.

§This can be produced from methanol, sodium bromide, and sulfuric acid in accord with the procedure in Reference 6, and liberated from admixed hydrogen bromide by passing it through a column with potassium hydroxide. The apparatus is first filled with nitrogen, and the liquid reagents are deaerated.

the reaction mixture for five hours. Since the *sec*-alkylcobalt complexes are especially sensitive to protonolysis, they should be crystallized from either alkalized or aprotic (e.g., $CHCl_3$) solvents.

Properties

1. The ethylcobalt complex crystallizes well as red prisms from methanol (or its mixtures with water); therefore, homogeneity of a given sample can be checked by examination under a microscope. The crystals are moderately photosensitive; in the dark they can be stored for at least a year (at room temperature and in air). The complex is soluble, but moderately so, in polar organic solvents such as methanol, formamide, or dimethyl sulfoxide, and somewhat less soluble in water. In solution, the complex is more sensitive to light than it is in the solid state. The structure of the complex has been established by an X-ray single-crystal diffraction study.[1]

IR spectrum (KBr pellet): 3320 (m), 3295 (m-s), 3230 (vs, br), 3130 (s), 3065 (vw), 3030 (vw), 2970 (m), 2945 (m), 2895 (m), 2860 (m), 1600 (vs), 1580 (sh), 1540 (m), 1470 (m-w), 1445 (s), 1395 (vw), 1340 (s), 1275 (w), 1260 (w), 1240 (m), 1175 (sh), 1165 (m), 1150 (m), 1140 (m), 1110 (vw), 1085 (w), 1075 (m-w), 1050 (s), 1030 (m), 990 (w), 960 (vw), 935 (vw), 880 (m-w), 850 (w), 750 (m), 740 (s), 630 (w), 610 (vw), 585 (w), 565 (w), 540 (w), 530 (w), 505 (vw), 470 (w), 450 (vw), 420 (sh), 410 (w) cm^{-1}.

The UV-visible spectrum (in MeOH): λ ($10^{-3}\epsilon$): 331 nm (3.37), 373 nm (2.34), 430 nm (1.45).

The ion exchange TLC on SiO_2 (Czechoslovakian Silufol plates) with 0.1 *N* solutions of NaOAc in (a) MeOH-H_2O 4:1 (v/v): R_f = 0.32; (b) $HCONH_2$-H_2O 1:1 (v/v): R_f = 0.59.

2. The other complexes of the series are characterized like the ethyl homolog (except for X-ray data), and their properties are similar to those of the ethyl complex.

B. ALKYL[2-[1-[2-AMINOETHYL)IMINO]ETHYL]-PHENOLATO](1,2-ETHANEDIAMINE)COBALT(III) IODIDE (The Series of the [**1a**]I Complexes)

Procedures

■ **Caution.** *See caution in Section A.*

1. The Ethylcobalt Complex (R = Et)

The synthesis is carried out as in Section A, through addition of alkali. Then, after five minutes, the solution of sodium tetrahydroborate(1-) (0.75 g, 20 mmoles) in 5 mL of 5% aqueous sodium hydroxide (85 mmoles of pure NaOH) is introduced, followed by 0.25 mL of a 2% solution of palladium(II) chloride in 1 *M* aqueous potassium chloride. After 15 minutes, the water bath is removed, and the flask is shielded from light. Next, ethyl iodide (5 mL, 9.6 g, 65 mmoles) is added dropwise over 15 minutes, and stirring is continued for four hours.

The dark-brown suspension resulting from the synthesis is filtered through a glass filter of medium porosity. The precipitate (I) is washed consecutively with cold water, a small amount of acetone, and diethyl ether.

Then, it is extracted with dichloromethane until the solvent is no longer green.* The combined methanol/water/acetone liquor is concentrated under vacuum to a volume of 35 mL, and is then diluted with 20 mL of water; a saturated aqueous solution of sodium iodide (5 mL) is then added. The resulting solution is concentrated in the same manner to a volume of 25 mL. The resulting precipitate (II) is collected by filtration† and treated like precipitate I. Then both precipitates are combined, which yields 3.15 g (7.0 mmoles, 49%)‡ of brownish-red crystalline material. The product is recrystallized from methanol as dark-red prisms.

2. Other Complexes of the Series ([**1a**]I)

Related iodide salts with other primary and secondary alkyl ligands (R) can be obtained similarly and in comparable yields. Thus, the complexes with R = Me, *n*-Bu and *i*-Pr are prepared starting from the corresponding alkyl iodides.

With regard to crystallization of the *sec*-alkyl cobalt complexes, see Section A-2.

Properties

The characteristics of the iodide salts are either identical (cation exchange chromatographic behavior) or close (IR and electronic spectra, solubilities) to those of the related bromides (Section A).

*See Footnote ‡ in Section A, p. 166. The bath temperature should be kept below 40°.

†The liquor and the water extract contain the unipositive bis-*mer*-[N-(2-aminoethyl)-7-methylsalicylideneiminato]cobalt(III) cation complex, $\{Co^{III}[o\text{-}OC_6H_4C(Me){=}N(CH_2)_2NH_2]_2\}^+$. To isolate its iodide salt, the aqueous solution is concentrated under vacuum to a volume of 10 mL. After some hours, a solid precipitates. It is collected by filtration, dried by suction, and is then extracted with dichloromethane. The liquor is evaporated to dryness under vacuum, which yields the complex in question (as the monohydrate 3.68 g, 6.8 mmoles, 49%). It can be converted to the desired product ([**1a**)I) by treatment with $Na[BH_4]$/Pd and EtI in alkaline methanol solution under conditions similar to those of the basic procedure.

‡This yield can be increased: see the preceding footnote.

C. ALKYL[2-[1-[(2-AMINOETHYL)IMINO]ETHYL]PHENOLATO]-(1,2-ETHANEDIAMINE)COBALT(III) PERCHLORATE

(The Series of the [**1a**] ClO_4^- Complexes)

■ **Caution.** *Perchlorates are explosive and must be treated with great care. Heat must be avoided, and the reaction must be properly shielded.*

Procedure

To obtain a perchlorate of this series, a methanolic solution of the related bromide salt is mixed with an excess of a concentrated aqueous solution of sodium perchlorate. An orange, crystalline precipitate is collected by filtration, washed with water, and dried by suction.

Properties

Compared with IR spectra of the halide salts, those of the perchlorates exhibit the following features. a. There are additional bands characteristic of the perchlorate ion. b. The strong and broad bands (at ~3230 and 3130 cm^{-1}), attributed to the vibrations of N—H bonds of the bridge complex cations with halide anions, are lost, while sharp peaks at higher frequencies, related to N—H links with protons that do not participate in hydrogen bonding, grow in number. These features are exemplified with fragments of the IR spectra of [**1a**]ClO_4 with R = Et: a. ClO_4^- bands (Nujol mull): ν_3 = 1100 cm^{-1} (vs), ν_4 = 655 cm^{-1} (s-m). b. Stretching region of coordinated NH_2 groups, that is, the 3100–3400 cm^{-1} range (fluorolub mull): 3380 (w), 3350 (m), 3340 (s), 3310 (m), 3285 (s), 3180 (vw) cm^{-1}.

D. ALKYL[2-[1-[(3-AMINOPROPYL)IMINO]ETHYL]-PHENOLATO](1,3-PROPANEDIAMINE)COBALT(III) IODIDE (The Series of the [1b]I Complexes)

Procedures

■ **Caution.** *See caution in Section A.*

1. The ethylcobalt complex (R = Et) This synthesis deviates from that in Section B in two ways. a. The 1,2-ethanediamine is substituted by 1,3-

propanediamine* (2.3 mL, 2.05 g, 28 mmoles). b. An extra reagent, namely sodium iodide (~4.5 g, 30 mmoles), is added after dissolution of the cobalt(II) salt.

The dark-brown suspension resulting from the synthesis is filtered through a glass filter of medium porosity. The solid† is washed consecutively with cold water, a small amount of acetone, and diethyl ether. Next, it is extracted with dichloromethane until the solvent is no longer green.‡ The remaining solid is dried by suction, and 4.6 g (9.6 mmoles, 69%) of red-brown, prismatic crystals are obtained.

2. The methylcobalt complex (R = Me). It is prepared similarly, but with a lower yield (30%).

Properties

The solubilities of both complexes are markedly lower than those of the related complexes (of the n = 2 series) in all solvents tested (H_2O, MeOH, Me_2CO, CH_2Cl_2, $CHCl_3$). The IR spectrum of the ethylcobalt complex is similar to that of the corresponding complex of the latter series. The UV-visible spectrum of the ethylcobalt complex (in MeOH): λ ($10^{-3}\epsilon$); 373 nm (3.63); a broad absorption is also found over the same wavelength range as for the related complex of the n = 2 series.

References

1. I. Ya. Levitin, A. L. Sigan, E. K. Kazarina, G. G. Alexandrov, Yu. T. Struchkov, and M. E. Vol'pin, *J. Chem. Soc. Chem. Comm.*, 441 (1981).
2. I. Ya. Levitin, A. L. Sigan, R. M. Bodnar, R. G. Gasanov, and M. E. Vol'pin, *Inorg. Chim. Acta*, **76,** L169 (1983).
3. A. Bigotto, G. Costa, G. Mestroni, G. Pellizer, A. Puxeddu, E. Reisenhofer, L. Stefani, and G. Tauzher, *Inorg. Chim. Acta Rev.*, **4,** 41 (1970).
4. G. N. Schrauzer, J. W. Sibert, and R. J. Windgassen, *J. Am. Chem. Soc.*, **90,** 6681 (1968).
5. T. S. Roche and J. F. Endicott, *Inorg. Chem.*, **13,** 1575 (1974).
6. N. Weiner, *Org. Syn., Coll. Vol.*, **2,** 280 (1943).
7. A. E. Miller and W. B. Housman, Jr., *French Patent* No. 1,355,309/1964, *Chem. Abstr.*, **61,** 4214e (1964).

*This reagent can be synthesized in a single-step from acrylonitrile, ammonia, and hydrogen over Raney nickel.[7] The normal boiling point of the compound (135–136°) can be employed as a quality check.

†In some preparations it incorporates a green, powdery material that can be removed by fractional sedimentation of a water suspension of the solid.

‡See Footnote ‡ in Section A.

33. COMPLEXES OF 1,3-DIHYDRO-1-METHYL-2*H*-IMIDAZOLE-2-THIONE WITH COBALT(II)

Submitted by E. S. RAPER* and J. R. CREIGHTON*
Checked by B. R. FLYNN† and M. NORTON†

$$Co(NO_3)_2 + 2(4)\ \text{mimt} \longrightarrow [Co(\text{mimt})_{2(4)}(NO_3)_2]$$

Introduction

The S,N-containing heterocyclic molecule 1,3-dihydro-1-methyl-2*H*-imidazole-2-thione (mimt), together with its thiol tautomer, have significant coordination potential which is likely to be susceptible to variations in pH and other reaction conditions.

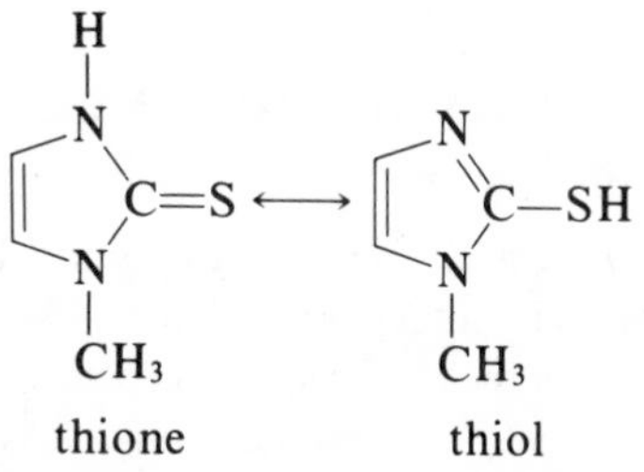

H–N, C=S, N–CH_3 ⟷ N, C–SH, N–CH_3

thione thiol

Reactions in neutral media have been reported to produce thione-sulfur donating complexes of mimt,[1,2] but it has been suggested that *cis*-$[PtCl_2(\text{mimt})_2]$, prepared under acid conditions, contains N-donating mimt.[3] The thione character of mimt has been demonstrated by infrared spectroscopy,[1,2] single crystal X-ray data, and ^1H nmr spectra.[4]

The imidazole-thione system possesses considerable pharmacological activity, with mimt being a particularly well-known and commercially available anti-thyroid (thyrotoxic) agent.[5]

Procedure

A. THE 1:4 (METAL:LIGAND) COMPLEX

A solution containing 0.45 g (4 mmol) of mimt* in 5 mL of a solution containing anhydrous ethanol and triethyl orthoformate (95:5% V/V) is added to a solution

*School of Chemical & Life Sciences, The Polytechnic, Newcastle-upon-Tyne, NE1 8ST, United Kingdom.
†Department of Chemistry, Broome Community College, Binghamton, NY 13902.

containing 0.29 g (1 mmole) of hydrated cobalt(II) nitrate ($Co(NO_3)_2 \cdot 6H_2O$) in 10 mL of the same solvent. The resultant mixture is refluxed in a single-necked, 50-mL, round-bottomed flask for 30 minutes. The total volume of solvent is then reduced slightly and the solution is allowed to cool to room temperature. Emerald-green crystals separate from this solution which are removed by suction filtration, washed with cold anhydrous ethanol (two × 5 mL) and vacuum dried at room temperature. Yield, 60% as $Co(C_4H_6N_2S)_4(NO_3)_2 \cdot H_2O$.

Properties

Tetrakis(1,3-dihydro-1-methyl-2*H*-imidazole-2-thione)dinitratocobalt(II) monohydrate is an emerald-green, crystalline solid. The Co—S contact is characterized by perturbation of the thioamide IV band of the parent ligand (730–745 cm^{-1}) and by a ν(Co—S) band at 310 cm^{-1}. The complex is non-conducting in nitromethane.

B. THE 1:2 (METAL:LIGAND) COMPLEX

A solution containing 0.22 g (2 mmoles) of mimt* in 5 mL of a solution containing ethyl acetate and triethyl orthoformate (95.5% v/v) is added to a solution containing 0.29 g (1 mmole) of hydrated cobalt(II) nitrate ($Co(NO_3)_2 \cdot 6H_2O$) in 10 mL of the same solvent. The resultant mixture is then refluxed in a 50-mL, single-necked, round-bottomed flask for 30 minutes. The subsequent treatment of the deep-blue crystalline product is the same as in (a). Yield, 90% as $Co(C_4H_6N_2S)_2(NO_3)_2$.

Properties

Bis(1,3-dihydro-1-methyl-2*H*-imidazole-2-thione)dinitratocobalt(II) is a deep-blue, crystalline solid. The Co—S contact is characterized by perturbation of the thioamide (IV) band of the parent ligand (730–740 cm^{-1}) and by a ν(Co—S) band at 315 cm^{-1}. The complex is non-conducting in nitromethane.

References

1. E. S. Raper and J. L. Brooks, *J. Inorg. Nucl. Chem.*, **39,** 2163 (1977).
2. E. S. Raper and I. W. Nowell, *Inorg. Chim. Acta,* **43,** 165 (1980).
3. J. Dehand and J. Jordonov, *Inorg. Chim. Acta,* **17,** 37 (1976).

*Supplied as the thiol tautomer by the Aldrich Chemical Company, Gillingham, Dorset, SP8 4JL, United Kingdom.

4. E. S. Raper, M. E. O'Neill, and J. A. Daniels, *Inorg. Chim. Acta*, **41(2)**, 145 (1980).
5. *Martindale Extra Pharmacopoeia*, 27th Edition, Pharmaceutical Press, London, 1977, p. 300.

34. METAL ION-CONTROLLED SYNTHESIS AND RING CONTRACTION OF A CONJUGATED, PLANAR, HEXADENTATE, MACROCYCLIC LIGAND

Submitted by S. MARTIN NELSON,* FERIDA S. ESHO,* JOÃO DE O. CABRAL,† M. FERNANDA CABRAL,† and MICHAEL G. B. DREW‡
Checked by COLIN J. CAIRNS§ and KIM LANCE§

An important aspect of the chemistry of macrocyclic ligands of limited flexibility relates to the structural/chemical interplay between the geometrical preferences of the ligand on one hand, and those of the complexed metal ion on the other. The following preparations illustrate the structural and chemical consequences of a change in metal ion size from "too large" to "too small" in relation to the fixed hole-size of a planar, conjugated macrocycle.

The macrocycle L^1 [structure (**1**)] having an inner ring of 18-member atoms has been synthesized (as its metal complexes) by the use of large metal ions as templates for the [2 + 2] Schiff base condensation of two molecules of 2,6-pyridinedicarboxaldehyde with two molecules of 1,2-benzenediamine.[2] The template action of the metal ion is evidenced by the observation that only oils or gums are obtained in metal-free reactions. When the Ba^{2+} ion (ionic diameter 2.84 Å)[2] is employed as template, the product complex has a 2:1 ligand:metal stoichiometry (Part A), whereas, for the Ca^{2+}, Sr^{2+} or Pb^{2+} ions (ionic diameter 2.24, 2.52, and 2.58 Å, respectively), the products have a 1:1 metal-to-ligand stoichiometry. Smaller metal ions are ineffective as templates for the synthesis of L^1.

However, it has been found[1,3] that once the macrocycle has been formed (using, for example, Ba^{2+} as a template), the template ion may often be replaced by another metal ion which is itself ineffective as a template. For the case where the incoming metal ion is Cd^{2+}, which is only slightly smaller (ionic diameter 2.20 Å) than Ca^{2+}, the replacement is effected without alteration in the macrocycle (Section B).[1] In contrast, where the incoming metal ion is appreciably smaller, as in the case of several first-row divalent transition metal ions (ionic

*Department of Chemistry, Queen's University, Belfast BT9 5AG, Northern Ireland.
†Laboratorio Ferreira da Silva, Faculdade de Ciencias, 4000 Porto, Portugal.
‡Department of Chemistry, The University, Reading RG6 2AD, United Kingdom.
§Department of Chemistry, The Ohio State University, Columbus, OH 43210.

diameters 1.80–1.92 Å),[2] the exchange in methanol solution is accompanied by an intramolecular contraction of the (hexadentate) 18-membered inner ring to a (pentadentate) 15-membered inner ring to afford a new macrocycle L^2 [structure **(2)**] of appropriate hole-size for the smaller metal ion (Section C).[3] X-ray analysis of cadmium(II) and lead(II) complexes of L^1 has shown[1] that the L^1 macrocycle has a near-planar conformation with an inner ring hole radius of about 2.7 Å. The structural and chemical results can thus be understood in terms of the relative sizes of the metal ion and L^1 macrocycle hole.

A. BIS(7,11:20,24-DINITRILODIBENZO[*b,m*][1,4,12,15]TETRA-AZACYCLODOCOSINE)BARIUM(II) PERCHLORATE

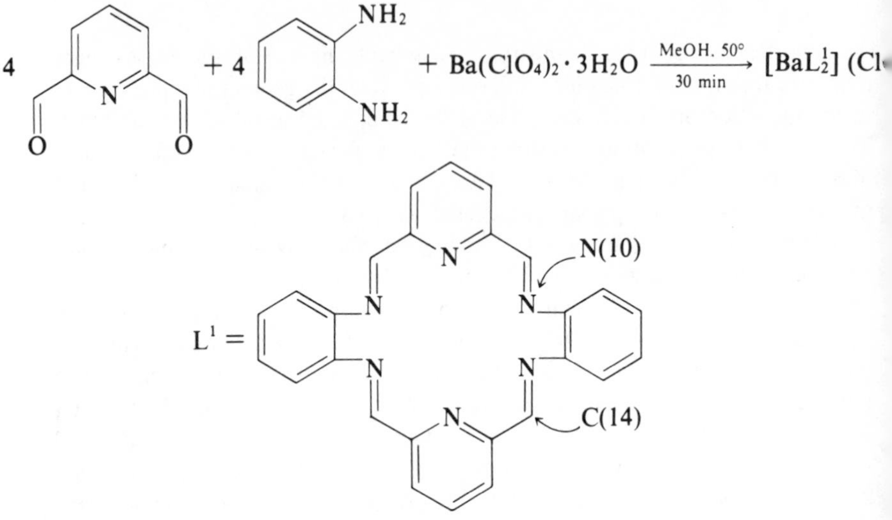

$$4\ (\text{2,6-pyridinedicarboxaldehyde}) + 4\ (\text{1,2-benzenediamine}) + Ba(ClO_4)_2 \cdot 3H_2O \xrightarrow[30\ \text{min}]{\text{MeOH, } 50^\circ} [BaL^1_2](Cl$$

L^1 = N(10) C(14)

(1)

Procedure

■ **Caution.** *Perchlorate salts of complexes can be explosive and should be handled with care in a fume hood and behind a protective shield. They should not be heated as solids or as concentrated solutions.*

A solution of barium perchlorate trihydrate (1.96 g, 0.005 mole) in 300 mL of methanol is placed in a 500-mL conical flask and heated to 50°. To this warm solution is added 1.37 g (0.01 mole) of freshly sublimed 2,6-pyridinedicarboxaldehyde,* followed by 1.20 g (0.01 mole) of 1,2-benzenediamine (after puri-

*The 2,6-pyridinedicarboxaldehyde may be prepared from commercially available 2,6-pyridinedimethanol by the method of E. P. Papadopoulos, A. Jarrar and C. H. Issidorides, *J. Org. Chem.*, **31**, 615 (1966).

fication with decolorizing charcoal and recrystallization from water). The mixture is stirred continuously and maintained at 50°. The initial yellow color of the solution deepens to yellow-orange, and, within minutes, an orange solid starts to separate. The temperature is maintained at ~50° and the stirring is continued for 30 minutes, after which time the mixture is cooled, the orange solid filtered off, washed with cold methanol, and dried in air or in a vacuum desiccator. A further batch of product may be recovered from the filtrate after standing overnight. Yield: 62%. *Anal.* Calcd. for $C_{52}H_{36}O_8N_{12}Cl_2Ba$: C, 53.59; H, 3.11; N, 14.43. Found: C, 53.43, H, 3.10; N, 14.58.

Properties

The $[BaL_2^1][ClO_4]_2$ is an orange, microcrystalline solid which is soluble in several polar organic solvents, such as acetonitrile, from which it may be recrystallized to give large, orange cubes. The infrared spectrum shows (among other things) bands at 1640 ($\nu_{C=N}$) 1400–1600 (pyridine and phenyl ring vibrations), 1090, and 620 cm^{-1} (non-coordinated ClO_4^-). No absorption occurs at 3200–3400 or ~1700 cm^{-1}, confirming the absence of the amino and carbonyl groups of the starting materials. The electrical conductance of 285 $\Omega^{-1}cm^2mole^{-1}$, found for a 10^{-3} *M* solution in acetonitrile at 25° is consistent with uni-bivalent electrolyte behavior. The ^1H nmr spectrum in dimethyl sulfoxide-d_6 shows a multiplet at 7.04–7.21 (phenyl protons), a doublet at 7.55, a triplet at 7.90 (pyridine protons), and a singlet at 8.36 ppm (imino protons). Integrated relative intensities for the three sets of resonances are in the required ratio of 4:6:2. Crystals of the complex belong to the $P\bar{4}m2$ space group with one molecule per unit cell, consistent with a "sandwich" structure, with the two planar macrocycles in a staggered relationship.

B. AQUA(7,11:20,24-DINITRILODIBENZO[*b,m*][1,4,12,15]TETRA-AZACYCLODOCOSINE-PERCHLORATOCADMIUM(II) PERCHLORATE·MeOH

$$[BaL_2^1][ClO_4]_2 + Cd(ClO_4)_2 \cdot 6H_2O \xrightarrow[6\ hr]{MeOH,\ 65^\circ} [CdL^1(H_2O)(ClO_4)](ClO_4) \cdot MeOH$$

Procedure

■ **Caution.** *See Section A.*

To a suspension of 1.17 g (0.001 mole) of $[BaL_2^1](ClO_4)_2$ in 500 mL of methanol in a 1-L round bottomed flask fitted with a reflux condenser is added 1.67 g (0.004 mole) of solid cadmium(II) perchlorate hexahydrate. The mixture is gently refluxed for 1 hour longer than it takes the Ba(II) complex to dissolve fully. The total reflux period is five to six hours. The resulting orange solution is concentrated to about 100 mL on a rotary evaporator and then set aside until the orange product crystallizes. The crystals are washed with cold methanol and dried in air. Yield: 65%. *Anal.* Calcd. for $C_{27}H_{24}O_{10}N_6Cl_2Cd$: C, 41.80; H, 3.12; N, 10.83. Found: C, 42.24; H, 3.03; N, 11.01.

Properties

The crystalline, orange complex is soluble to varying degrees in a variety of polar organic solvents. It may be recrystallized from acetonitrile. The infrared spectrum is very similar to that of $[BaL_2^1](ClO_4)_2$, except for the appearance of a broad absorption [$\nu_{(O-H)}$] at ~3500 cm^{-1}, due to the H_2O and CH_3OH molecules, and a broadening of the ClO_4^- absorptions at ~1090 cm^{-1} and 620 cm^{-1}, reflecting coordination of one of these anions in the solid state. The complex is a uni-bivalent electrolyte (Λ_M = 286 $\Omega^{-1}cm^2mole^{-1}$ for a $10^{-3}M$ solution) in acetonitrile at 25°. The ^1H nmr spectrum in dimethyl sulfoxide-d_6 is also very similar to that of $[BaL_2^1](ClO_4)_2$. The coordination geometry about the Cd(II) atom is approximately hexagonal bipyramidal, with the six nitrogen atoms of L^1 defining an approximate hexagonal plane (mean Cd-N distance 2.63 Å), and with the oxygen atoms of a water molecule and a perchlorate ion occupying the two axial positions.[1]

C. COBALT(II) COMPLEX OF THE RING CONTRACTED MACROCYCLE

$$[BaL_2^1](ClO_4)_2 + 2\ Co(ClO_4)_2\cdot 6H_2O \xrightarrow[\text{3 days}]{\text{MeOH, 65°}} [CoL^2(H_2O)(MeOH)](ClO_4)_2$$

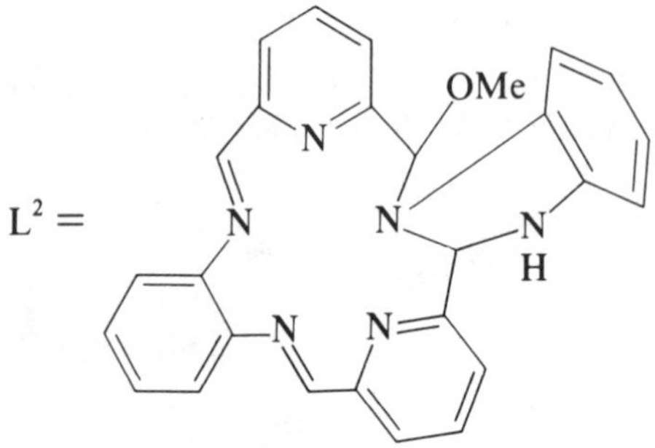

L^2 =

(2)

5,5a-dihydro-24-methoxy-6,10:19,23-dinitrilo-24*H*-benzimidazo[2,1-h][1,9,17]benzotriazacyclononadecine

Procedure

■ **Caution.** *See Section A.*

To a suspension of 1.17 g (0.001 mole) of $[BaL_2^1](ClO_4)_2$ in 800 mL of methanol, in a 2-L round-bottomed flask fitted with a reflux condenser, is added 2.2 g (0.006 mole) of solid cobalt(II) perchlorate hexahydrate. The mixture is gently refluxed until the Ba(II) complex has completely dissolved, which may take two to three days. After the solution becomes clear, the reflux should be continued for an additional two hours. The orange solution is then concentrated to a volume of about 50 mL by rotary evaporation and then set aside in a beaker to crystallize. The orange crystals which separate are filtered, washed with cold methanol, and dried in air. Yield: 60%. *Anal.* Calcd. for $C_{28}H_{28}O_{11}N_6Cl_2Co$: C, 44.60; H, 3.74; N, 11.15. Found: C, 44.26; H, 3.61; N, 11.00.

Properties

The orange, crystalline complex is readily soluble in a range of polar organic solvents including methanol, acetonitrile, and dimethyl sulfoxide. The infrared spectrum differs from those of the L^1 series of complexes in showing a ν_{N-H} band at 3370 cm^{-1} and a $\nu_{C=N}$ vibration of reduced intensity at 1615 cm^{-1}. The magnetic moment of the solid is 5.01 BM at 293 K, corresponding to a high-spin ($S = 3/2$) ground state for the metal ion. As shown in structure **(2)** the inner large ring of L^1 has contracted from 18- to 15-member atoms, as result of covalent bond formation between the N(10) and C(14) of L^1, with accompanying formation of a five-membered imidazolidine ring.[3] The ring contraction has left the Co(II) atom in a pentagonal bipyramidal environment, with the five nitrogen atoms of the 15-membered inner large ring making up the equatorial plane, the two axial sites being occupied by the oxygen atoms of a water and a methanol molecule.

The ring contraction can be seen as proceeding in two steps. The first is the addition of a molecule of solvent methanol across one C=N bond of L^1, as a result of the strain generated by the attempts of the macrocycle to bond effectively by way of all six nitrogen atoms to a "too-small" metal ion. The gain in flexibility accompanying the conversion of a double into a single carbon-nitrogen bond now allows nucleophilic addition of the secondary NH group across the neighboring azomethine bond to form L^2.

Analogous metal-promoted ring contractions of L^1 to L^2 may be carried out using salts of Mn(II), Fe(II), and Zn(II) in place of Co(II).[3]

Acknowledgement

We thank the Department of Scientific Affairs, N.A.T.O., for support. J.O.C. and M.F.C. also thank I.N.I.C. (Lisbon).

References

1. M. G. B. Drew, J. de O. Cabral, M. F. Cabral, F. S. Esho, and S. M. Nelson, *Chem. Commun.*, 1033 (1979).
2. R. D. Shannon, *Acta Cryst.*, **A32,** 751 (1976). The values quoted in the text refer to the octacoordinate ions.
3. S. M. Nelson, F. S. Esho, M. G. B. Drew, and P. Bird, *Chem. Commun.*, 1035 (1979).

35. TRIPHENYLPHOSPHINE OXIDE AND BIPYRIDINE DIOXIDE COMPLEXES OF CERIUM(IV)

Submitted by FRANTIŠEK BŘEZINA*
Checked by R. P. PERITO† and B. B. CORDEN†

Cerium(IV) forms double salts readily, the best-known one being cerium(IV) ammonium nitrate, $Ce(NO_3)_4 \cdot 2NH_4NO_3$. Although many complex cationic and anionic cerium(IV) species are present in solution, few have been isolated, and these are mostly with anions as ligands. Since the 4f transition elements often form complexes with oxygen compounds, it is possible to prepare Ce(IV) complexes with phosphine oxides or 2,2′-bipyridine 1,1′-dioxide.[1–3]

A. TETRANITRATOBIS(TRIPHENYLPHOSPHINE OXIDE)CERIUM(IV)

$$(NH_4)_2[Ce(NO_3)_6] + 2\ OPPh_3 \longrightarrow [Ce(NO_3)_4(OPPh_3)_2] + 2\ NH_4NO_3$$

Procedure

Ammonium hexanitratocerate(IV) (2.0 g, 3.64 mmoles) in acetone (20 mL) is treated with triphenylphosphine oxide, $OPPh_3$ (2.05 g, 7.36 mmoles), in acetone (20 mL). The mixture is filtered to remove a white precipitate (NH_4NO_3), and the solvent is removed under reduced pressure. The orange crystals are washed with chloroform and recrystallized from acetonitrile. *Anal.* Calcd. for $Ce(NO_3)_4 \cdot 2\ OPPh_3$: Ce, 14.84; NO_3^-, 26.26; P, 6.56. Found: Ce, 14.76; NO_3^-, 26.20; P, 6.40.

*Palacky University, Olomouc, Czechoslovakia.
†Department of Chemistry, Tufts University, Medford, MA 02155.

B. TETRACHLOROBIS(TRIPHENYLPHOSPHINE OXIDE)CERIUM(IV)

$$H_2[CeCl_6] + 2\ OPPh_3 \longrightarrow [CeCl_4(OPPh_3)_2] + 2\ HCl$$

Procedure

To a solution of dihydrogen hexachlorocerate(IV), prepared[4] from (1.1 g, 2.0 mmoles) of $(NH_4)_2[Ce(NO_3)_6]$, in 10 mL of dry methanol, is added a solution of triphenylphosphine oxide (1.1 g, 4.0 mmoles) in 5 mL of dry methanol. A stream of dry hydrogen chloride gas is bubbled through this reaction mixture at −20° for a half hour. The cerium(IV) complex precipitates, and the orange product is filtered and is washed with dry methanol and petroleum ether. *Anal.* Calcd. for $CeCl_4 \cdot 2\ OPPh_3$: Ce, 16.72; Cl, 16.92; P, 7.32. Found: Ce, 16.71; Cl, 16.70; P, 7.19.

C. TETRAKIS(2,2′-BIPYRIDINE 1,1′-DIOXIDE)CERIUM(IV) NITRATE

$$(NH_4)_2[Ce(NO_3)_6] + 4\ bpyO_2 \longrightarrow [Ce(bpyO_2)_4](NO_3)_4 + 2\ NH_4NO_3$$

Procedure

To a warm solution of 2,2′-bipyridine 1,1′-dioxide (1.6 g, 9.3 mmoles) in 20 mL of water is added ammonium hexanitratocerate(IV) (1.1 g, 3.95 mmoles). The reaction mixture is kept at approximately 95° in a water bath for a half hour. Then an excess of $NH_4[PF_6]$ is added to effect precipitation of the complex as a yellow-orange crystalline solid, which is washed with ethanol and dried. *Anal.* Calcd. for $[Ce(bpyO_2)_4][PF_6]_4$: C, 32.63; H, 2.19; N, 7.61; P, 8.41. Found: C, 32.08; H, 1.99; N, 7.70; P, 7.94.

References

1. F. Březina, *Coll. Czech. Chem. Commun.*, **39,** 2162 (1974).
2. Mazhar Ul-Haque, C. N. Caughlan, F. A. Hart, and R. Van Nice, *Inorg. Chem.*, **10,** 115 (1971).
3. F. Březina, *Coll. Czech. Chem. Commun.*, **36,** 2889 (1971).
4. S. S. Moosath and M. R. A. Rao, *Proc. Ind. Acad. Sci.*, **43 A,** 213 (1956).

36. HEXAKIS(DIPHENYLPHOSPHINIC AMIDE)LANTHANOID(III) HEXAFLUORO PHOSPHATES, $[Ln(dppa)_6](PF_6)_3$*

$$[Ln(PF_6)_3]\cdot xH_2O + 6\ dppa \xrightarrow[\text{MeOH and 2,2-dimethoxypropane}]{\text{EtOH or}} [Ln(dppa)_6](PF_6)_3 + x\ H_2O$$

Submitted by G. VICENTINI† and L. B. ZINNER†
Checked by S. C. KUO‡ and L. THOMPSON‡

The hexafluorophosphate ion has proved to be very useful for studies with complexes, since the $[PF_6]^-$ anion has a very weak coordinating ability.[1] Lanthanoid hexafluorophosphates can be obtained in very concentrated aqueous solutions by reaction of a freshly prepared solutions of hexafluorophosphoric acid and a hydrated lanthanoid basic carbonate.[2] The resulting solution, after filtration, is evaporated to near dryness. Attempts to isolate the hydrated salts are unsuccessful because of decomposition accompanied by hydrogen fluoride evolution. Nevertheless, the complexes containing diphenylphosphinic amide are isolable and are quite stable.[3]

Preparation of Starting Materials

The lanthanoid basic carbonate is obtained by boiling a dilute lanthanoid chloride solution with urea.[4] The dilute solution of freshly prepared hexafluorophosphoric acid is obtained by percolating a solution, containing 3.0×10^{-3} mole of ammonium hexafluorophosphate (Alfa products) in 20 mL of water, through an Amberlite IR $-$ 120 H^+ column (1 cm $\times$ 25 cm). The eluted solution is allowed to fall dropwise (~20 drops per minute) onto a suspension of 1×10^{-3} mole of the lanthanoid basic carbonate in 5 mL of water, under constant stirring. The addition of the acid is discontinued when a very small residue of the basic carbonate remained ($pH > 5 < 6$). The solution is then filtered, the residue washed with two portions of 3 mL of distilled water, and then allowed to evaporate at ~30° to near dryness, using a flash evaporator§ and reduced pressure (~20 torr). It is necessary to evaporate carefully to a very small volume, because the solution decomposes, becoming acid and turbid. (This is a very critical point.)

*(Ln = La − Lu, ddpa = diphenylphosphinic amide).

†Instituto de Química, Universidade de São Paulo, Caixa Postal 20.780, São Paulo, Brazil.
‡Department of Chemistry, University of Minnesota-Duluth, Duluth, MN 55812.

§The checkers report that use of a rotary evaporator is satisfactory.

For the preparation of dppa, diphenylphosphinic acid (Aldrich) was transformed into diphenylphosphinic chloride by reaction with sulfinyl chloride[5] (82% yield). The dppa (mp 164–166°) was obtained in 75% yield by reaction of the diphenylphosphinic chloride with concentrated ammonia, according to the method of Zhmurova et al.[6]

Procedures

The complexes having the general formula $[Ln(dppa)_6][PF_6]_3$ are prepared in two ways: (a) the concentrated lanthanoid hexafluorophosphate solution is diluted with 5 mL of absolute ethanol and treated with a solution of dppa (1×10^{-2} mole in 20 mL of absolute ethanol). The precipitate is collected in a sintered-glass filter funnel, washed with two 5-mL portions of absolute ethanol, and dried under vacuum over anhydrous calcium chloride; (b) the concentrated aqueous solution of lanthanoid hexafluorophosphate is diluted with 5 mL of methanol and treated with a solution of dppa (1×10^{-2} mole in 10 mL of methanol), and the complex is then precipitated by addition of ~30 mL of 2,2-dimethoxypropane (Aldrich). The precipitate is collected, washed with two portions of 5 mL of 2,2-dimethoxypropane, and dried as above.[3]

Properties

The compounds are non-hygroscopic, quite stable,* melt at ~155°, and are colorless, except for those of praseodymium, neodymium, and erbium, which show green, blue, and pink hues, respectively. They are very soluble in acetonitrile, nitromethane, and methanol, and behave as 1:3 electrolytes in acetonitrile and nitromethane (millimolar solutions).

The IR spectra show no coordinated hexafluorophosphate ions, since only two bands (ν_3 ~840 and ν_4 ~560 cm^{-1}) are observed. Coordination through the phosphoryl oxygen is indicated by the shift of the P═O (~1130 cm^{-1}) stretching frequency to lower frequencies, and by the shift of the νP—N (~930 cm^{-1}) to higher frequencies, as compared to the free ligand (dppa νP═O, 1175 and νP—N, 915 cm^{-1}).

The X-ray powder patterns indicate that the entire lanthanoid series (except for promethium), and yttrium comprise an isomorphous series.[3]

The number of bands in the electronic spectrum of the neodymium compound at 77 K, due to the $^4I_{9/2} \rightarrow {}^4G_{5/2}, {}^2G_{7/2}$ transitions, are indicative of a cubic site around the central ion. The nephelauxetic parameter ($\bar{\beta} = 0.9917$), the covalent factor ($b^{1/2} = 0.064$), and the Sinha parameter ($\delta = 0.83$) show an almost

*For long-term storage, it is recommended that plastic flasks be used.

complete electrostatic character for the Ln^{3+}-dppa interaction. The oscillator strength in nitromethane solution is 9.3×10^{-6} cm^2 mole^{-1} L.[7]

The europium emission spectrum at 77 K suggests an $O_h \rightarrow C_{4v}$ symmetry and an octahedral geometry.[7] A single crystal X-ray study of the compound indicates a crystallographic octahedral symmetry with a = 20.28(1) Å, space group F23. The Ln-O distance is 2.27(1) Å.[8]

References

1. H. G. Mayfield, Jr. and W. E. Bull, *J. Chem. Soc. A*, 2279 (1971).
2. G. Vicentini, L. B. Zinner, and L. R. F. Carvalho, *J. Inorg. Nucl. Chem.*, **37,** 607 (1975).
3. L. B. Zinner and G. Vicentini, *Inorg. Chim. Acta*, **15,** 235 (1975).
4. G. Vicentini, J. M. V. Coutinho, and J. V. Valarelli, *An. Acad. brasil. Ciênc.*, **36,** 123 (1964).
5. (a) N. Kreutzkamp and H. Schindler, *Arch. Pharm.*, **293,** 296 (1960); (b) *Ibid., Chem. Abstr.*, **60,** 4179 (1964).
6. (a) I. N. Zhmurova, I. Yu. Voitsekhovskaya, and A. V. Kirsanov, *Zh. Obshch. Khim.*, **29,** 2083 (1959); (b) *Ibid., Chem. Abstr.*, **54,** 8681 (1960).
7. L. R. F. Carvalho, G. Vicentini, and K. Zinner, *J. Inorg. Nucl. Chem.*, **43,** 1088 (1981).
8. G. Oliva, E. E. Castellano, G. Vicentini, and L. R. F. Carvalho, *VI Symposium of the Science Academy of São Paulo State: Rare-Earth Chemistry*, 1981.

37. COMPLEXES OF THE TYPES: $K_3[Cr(NO)(CN)_5]$ AND $[Cr(NO)(NCS)_2(LL)]$ (LL = 2,2′-BIPYRIDINE; 1,10-PHENANTHROLINE

Submitted by RAMGOPAL BHATTACHARYYA,* GOBINDA P. BHATTACHARJEE,* PARTHA S. ROY,* and NABANITA GHOSH*
Checked by R. DEL ROSARIO,† S. C. BLAKE,† S. J. CARTER,† and L. S. STUHL†

The known[1-3] reductive nitrosylation reaction of tetraoxometallates using NH_2OH in alkaline medium has been little used in inorganic syntheses.[3] Recently, it has been shown that this reaction also occurs in neutral and slightly acidic media.[4] However, those syntheses succeeded in only partially converting the oxoanions (especially for CrO_4^{2-} and VO_4^{3-}) to the lower valent metal-nitrosyl derivatives, as is evident from the low yields of the products isolated,[1-5] and from the failure to isolate 2,2′-bipyridine (bpy) and 1,10-phenanthroline (phen) derivatives from the reaction mixture, even in the case of molybdenum.[5]

*Department of Chemistry, Jadavpur University, Calcutta 700 032, India.
†Department of Chemistry, Brandeis University, Waltham, MA 02254.

Moreover, until very recently,[6] the reductive nitrosylation of CrO_4^{2-}, both in alkaline[2] and acidic media,[4a] gives a major amount of unconverted residue.[7] The previous workers[1–5] used only 3 or 4 moles of nitrosylating agent (NH_2OH) per mole of CrO_4^{2-}, and this small ratio gave a poor yield of the metal-nitrosyl derivatives. It has recently been discovered[6] that a significant improvement of the efficiency of the nitrosylating process in a slightly acidic medium (pH = 5) can be observed if an excess of $NH_2OH \cdot HCl$ (10–15 moles per mole of the oxoanion) is used in the presence of an excess of the NCS^- ion. The same is true for the reductive nitrosylation of CrO_4^{2-} in a strong alkaline medium in the presence of the CN^- ion.

A. (2,2′-BIPYRIDINE)NITROSYL-BIS(THIOCYANATO)CHROMIUM(I)

Procedure

A solution of 1.0 g (0.0052 mole) of K_2CrO_4 and 3.9 g (0.052 mole) of NH_4SCN in 40 mL H_2O, is placed in a 250-mL beaker, 5.4 g (0.078 mole) of $NH_2OH \cdot HCl$ is added, and the mixture is heated to 80° on a hot plate for about 40 minutes, with constant stirring by a magnetic bar. A greenish-violet solution is obtained, which is then cooled to 27° and filtered through a fine sintered-glass filter to remove precipitated sulfur. An aqueous (50 mL) bpy (1.22 g, 0.0078 mole) solution is warmed (60°) and then the greenish-violet* solution is poured dropwise into it over a period of five minutes, with vigorous stirring. After the addition, stirring is continued for an hour with occasional scratching, while keeping the reaction mixture warm (50°). A yellowish-brown solid separates, after which the reaction mixture is cooled to the room temperature and filtered through a fine sintered-glass filter. The precipitate is washed thoroughly with cold water, a cold water–90% ethanol mixture (1:1), and then with diethyl ether, and is dried over $CaCl_2$ under vacuum. Yield of the crude product is 1.79 g (98%). The product is then extracted with acetone. Evaporation of the acetone yields bright, yellowish-brown crystals of $[Cr(NO)(NCS)_2bpy]$. A typical yield of the pure product is 1.50 g (82%).

The product is identified by its magnetic susceptibility (μ_{eff} = 2.0 BM at 27°), ESR (g_{av} = 1.98), infrared spectrum, and elemental analysis. *Anal.* Calcd. for $[Cr(NO)(NCS)_2C_{10}H_8N_2]$: Cr, 14.7; C, 40.7; H, 2.3; N, 19.8; S, 18.1. Found: Cr, 14.5; C, 41.0; H, 2.6; N, 19.6; S, 18.5. Infrared spectrum: ν_{NO} 1705; ν_{CN} 2080; ν_{CrN}(NO) 590; ν_{NCrN}(bpy) 328, 362 cm^{-1}.

*The checkers report the color to be brownish.

B. NITROSYL(1,10-PHENANTHROLINE) DITHIOCYANATOCHROMIUM(I)

A mixture of 1.0 g (0.0052 mole) of K_2CrO_4, 3.9 g (0.052 mole) of NH_4SCN, and 5.4 g (0.078 mole) of $NH_2OH \cdot HCl$ in 40 mL H_2O is treated as the bpy derivative of chromium(I) described above, to yield a greenish-violet* solution. Then, an aqueous solution of 1,10-phenanthroline (1.55 g, 0.0078 mole) is added and the mixture stirred as in the case of bpy analog. The crude product (yield 1.9 g, 98%) is then crystallized from acetone as described above, to obtain yellowish-brown crystals. Yield 1.63 g (84%).

The compound is characterized in the same manner as its bpy analog. *Anal.* Calcd. for $[Cr(NO)(NCS)_2(C_{12}H_8N_2)]$: Cr, 13.8; C, 44.4; H, 2.1; N, 18.5; S, 16.9%. Found: Cr, 13.4; C, 44.0; H, 2.5; N, 18.1; S, 17.0%. The magnetic susceptibility data, ESR spectrum, and the selected infrared bands are identical with those of the bpy analog.

Properties

The thiocyanato compounds are yellowish-brown with a luster. These complexes are stable in air for a considerable period (~30 days), and are soluble in acetone, acetonitrile, and N,N-dimethylformamide. The uv-visible spectra of the compounds in acetonitrile medium shows λ_{max} at 645, 490 (sh), and 350 nm (for both the bpy and phen compounds).

C. TRIPOTASSIUM PENTACYANONITROSYLCHROMATE(I) MONOHYDRATE

Procedure

A mixture of 1.0 g (0.0052 mole) K_2CrO_4, 4.02 g (0.0618 mole) of KCN, and 3.46 g (0.0618 mole) of KOH is dissolved in 25 mL of H_2O. It is then heated at 80° with stirring. Then, 3.58 g (0.052 mole) of $NH_2OH \cdot HCl$ is added in small portions. The stirring is continued for an hour while keeping the temperature of the solution at 80°. A yellowish-green solution is obtained, which is cooled below 10° and poured into 100 mL of cold (15°) ethanol (90%), dropwise and without any stirring. A green precipitate appears slowly. After completion of the addition, the resulting solution is kept cold (15°) for 20 minutes to allow the green precipitate to settle. The supernatant liquid is then carefully decanted to remove some of the white, flocculent mass suspended over the green precipitate.

*The checkers report a brownish color.

The latter is then collected by filtration through a fine sintered-glass filter and washed thoroughly with 40 mL of cold ethanol (90%, 15°) and then with diethyl ether.

The yellowish-green product is dissolved in 4 mL of water (filtered, if necessary), and the green solution is poured dropwise into 50 mL of cold ethanol (90%, 15°). The green precipitate is allowed to settle while still cold, and the alcohol is decanted as before to remove the white, flocculent mass. The precipitate is collected on a fine sintered-glass filter, washed thoroughly with cold ethanol, and then with diethyl ether, and dried under vacuum. This recrystallization is repeated once more by dissolving the product in 3.5 mL of H_2O and precipitating with 50 mL of cold (15°) ethanol (90%), whereupon bright green crystals are obtained. It is then washed with cold ethanol and diethyl ether, and then dried over fused $CaCl_2$ under vacuum. Yield: 0.750 g. (40%).

Note. The cooling operations described in the above procedures are necessary so as to cause the heavy nitrosyl complex to settle to the bottom of the beaker, while the light KCl, etc. will remain flocculent and settle over the former. It is then easy to decant a majority of the white, undesirable salts (KCl, KCN, etc.), so that the pure product can be obtained after only two crystallizations.

A preliminary test for purity (i.e., before the product is analyzed for the elements) can be conducted easily by measuring the magnetic susceptibility. A pure product (i.e., free from admixture of KCl, KCN, etc.) shows a μ_{eff} of 1.85–1.92 BM[2] at 27°. *Anal.* Calcd. for $K_3[Cr(NO)(CN)_5]\cdot H_2O$: C, 17.27; N, 24.18; H, 0.58; Cr, 14.97; K, 33.77. Found: C, 16.5; N, 23.7; H, 0.8; Cr, 15.07; K, 34.1. Infrared spectrum: ν_{NO} 1637, ν_{CN} 2120, ν_{CrN} (NO) 620, ν_{CrC} 420, 400, and δ_{CrCN} 340 cm^{-1}.[8]

Properties

The compound is identified by its magnetic susceptibility (μ_{eff} = 1.87 BM at 27°), ESR[9] and electronic spectral bands.[9] The λ_{max} wavelengths are at 730, 450, 365, 268, and 230 nm.

References

1. W. Hieber, R. Nast, and G. Gehring, *Z. anorg. allg. Chem.*, **256,** 169 (1948).
2. W. P. Griffith, J. Lewis, and G. Wilkinson, *J. Chem. Soc.*, 872 (1959).
3. K. G. Caulton, *Coord. Chem. Revs.*, **14,** 317 (1976).
4. (a) S. Sarkar and A. Muller, *Z. Naturforsch,* **33b,** 1053 (1978); (b) A. Muller, U. Sayer, and W. Eltzner, *Inorg. Chim. Acta,* **32,** L65 (1979).
5. S. Sarkar and P. Subranaaniam. *Inorg. Chim. Acta,* **35,** L357 (1979).
6. R. G. Bhattacharyya, G. P. Bhattacharjee, and P. S. Roy, *Inorg. Chim. Acta.*, **54,** L263 (1981).
7. R. G. Bhattacharyya and G. P. Bhattacharjee—Unpublished work.

8. K. Nakamoto, *Infrared and Raman Spectra of Inorganic and Coordination Compounds*, Wiley-Interscience, New York, 1978.
9. J. H. Enemark and R. D. Feltham, *Cord. Chem. Revs.*, **13**, 339 (1974).

38. HETEROPOLYTUNGSTATES WITH UNSATURATED HETEROPOLYANIONS

Submitted by G. H. MARCU* and A. BOTAR*
Checked by S. C. TERMES† and M. T. POPE†

Recently, new representative compounds of the heteropolyanions have been synthesized, which are characterized by the presence of two heteroatoms in the polytungstic anion. They are heteropolycompounds of the 1:1:11 and 1:2:17 types. These compounds are obtained from 1:12 or 2:18 heteropolyanions by a controlled alkaline degradation in the presence of one metallic cation, or by the direct reaction between unsaturated heteropolyanions obtained previously and a metallic cation.[1-8]

It is known that unsaturated heteropolyanions react easily with different di-, tri-, and tetravalent cations, forming very stable heteropolycompounds. Depending upon the nature of the cation, the stoichiometry of the metallic cation and the unsaturated heteropolyanion may be 1:1 or 1:2. More recently, a series of organic derivates of the heteropolyanions have been obtained, starting from heteropoly compounds belonging to the 1:2 class by a substitution reaction of the WO^{4+} unit with an organic cation, or by a direct reaction between an organic cation and an unsaturated heteropoly anion.[9-17]

A. POTASSIUM BIS(UNDECATUNGSTOPHOSPHATO)URANATE(IV)

$$U(SO_4)_2 + 2\ K_7[PW_{11}O_{39}] \longrightarrow K_{10}[U(PW_{11}O_{39})_2] + 2\ K_2SO_4$$

■ **Caution.** *Uranium is a radioactive element and it must be handled with particular care. Fume hoods and other apparatus appropriate for use with radioactive substances must be used. Waste solutions are a source of radioactive contamination, and they must be disposed of properly.*

*Institute of Chemistry, 65-103, Str. Donath, 3400 Cluj-Napoca, Roumania.
†Department of Chemistry, Georgetown University, Washington, DC 20057.

A solution containing uranium(IV) is prepared from 0.8 g (0.0016 mole) of $UO_2(CH_3COO)_2 \cdot 6H_2O$ dissolved in 10 mL water and reduced from U(VI) to U(IV).[3,5,17] This solution is added gradually to a lukewarm solution (70°) of 10 g (0.003 mole) of $K_7[PW_{11}O_{39}] \cdot 14H_2O$ in 40 mL of water. The pH of the solution is adjusted to 5.0 using glacial acetic acid. This pH of 4.5–5.0 is maintained during the addition by using a buffer containing potassium acetate in solution. After stirring for 15 minutes at a constant temperature of 70°, the dark brown solution is cooled to room temperature, filtered to eliminate possible traces of any precipitate, and is allowed to stand at 5°. After 2–3 days, dark brown crystals are separated from the solution, filtered through a filtering crucible, washed with small quantities of cool water, and dried in air in a dark room.[14] The yield is 3.2 g (30%). *Anal.* Calcd. for $K_{10}[U(PW_{11}O_{39})_2] \cdot 22H_2O$: K, 6.12; U, 3.72; W, 63.29; P, 0.97, H_2O, 6.15. Found: K, 6.18; U, 3.70; W, 63.40; P, 0.99; H_2O, 6.42.

Properties

The brown crystals of $K_{10}[U(PW_{11}O_{39})_2] \cdot 22H_2O$ are paramagnetic with $\mu_{eff} = 2.78$ BM. This indicates that the uranium is in the 4+ oxidation state. The synthesized U(IV) heteropolytungstate belongs to the class of U(IV) compounds with cubic symmetry. The visible absorption spectrum shows characteristic bands for octa-coordinated U(IV). The L band (14,814 cm^{-1}) is well resolved, but the O,R,U, and W bands (16,000; 20,010; 22,498; and 24,656 cm^{-1}, respectively) are weaker. The UV absorption spectrum is similar to that of $K_7[PW_{11}O_{39}] \cdot 14H_2O$, but it shows less easily distinguished maxima, which are shifted to lower energies, and the electron transitions are more evident in U(IV) heteropolytungstate than in the unsaturated heteropolyanion. The characteristic band at 38,910 cm^{-1} is attributed to the electron transitions of the tricentric bond W—O—W, and the band at 49,826 cm^{-1} is attributed to the p_π -d_π transition in the tungsten-oxygen bond.

The IR spectrum has three absorption bands in the 700–1000 cm^{-1} region (at 800, 890, and 965 cm^{-1}), due to the valence vibrations of the W═O and W—O—W bonds. The splitting of the 1080 cm^{-1} band (into 1060 and 1095 cm^{-1} bands) is attributed to stretching of the P—O bond. The rapid isotopic exchange recorded for the tungsten in U(IV) heteropolytungstate shows that W—O bonds in WO_6 octahedra are labile because of the partially covalent and partially ionic nature of this bond, and because of the unequivalence of the W—O bonds in WO_6 octahedra. The absence of the isotopic exchange of phosphorus shows that P—O bonds are inert for the exchange of the phosphorus, which is situated in the center of a regular tetrahedron. The X-ray diffraction establishes the fact that $K_{10}[U(PW_{11}O_{39})_2] \cdot 22H_2O$ has a cubic structure, namely it is a face-centered cube with a = 10.72 Å. Salts of other cations may be obtained by double exchange reactions in aqueous medium. The stability constant of the complex, determined by paper high voltage electrophoresis, is 2.48×10^6. The acid $H_{10}[U(PW_{11}O_{39})_2]$ is obtained in aqueous solution, and is stable for two weeks.

B. POTASSIUM BIS(HEPTADECATUNGSTODIPHOSPHATO)URANATE(IV)

$$2\ K_{10}[P_2W_{17}O_{61}] + U^{4+} \longrightarrow K_{16}[U(P_2W_{17}O_{61})_2] + 4\ K^+$$

■ **Caution.** *See Section A.*

A solution of uranium(IV) is prepared from 1.3 g (0.0025 mole) of $UO_2(CH_3COO)_2 \cdot 6H_2O$ dissolved in 20 mL of water, followed by reduction of U(VI) to U(IV).[3,5,17] This solution is gradually added (with continuous stirring) to a solution containing 27 g (0.005 mole) of $K_7[P_2W_{17}O_{61}] \cdot 22H_2O$ in 75 mL of water (60°). The pH of this solution is adjusted to 5.0 using glacial acetic acid. The stirring is continued for another 15 minutes, after which the dark violet solution is filtered to remove traces of any precipitate. The solution is allowed to stand at 5° for two to three days, whereupon violet microcrystals appear. These crystals are filtered through a filtering crucible, washed with small quantities of cool water, and dried in air in a dark room.[13]

The yield is 10.6 g (40%). The crude product is recrystallized from the minimum quantity of water at 60° at pH 4.5. *Anal.* Calcd. for $K_{16}[U(P_2W_{17}O_{61})_2] \cdot 22H_2O$: K, 6.36; U, 2.38; W, 63.60; P, 1.32; H_2O, 5.85. Found: K, 6.40; U, 2.42; W, 63.71; P, 1.36; H_2O, 5.90.

Properties

The dark violet crystals are paramagnetic; $\mu_{eff} = 2.98$ BM. The visible absorption spectrum shows absorption bands specific for the f^2 configuration of the octacoordinated U(IV); the best-resolved band is the L band at 14,812 cm^{-1}. The UV absorption spectrum shows a strong band at 38,000–40,000 cm^{-1} specific to the heteropolytungstates of the 2:18 class, and another band at 49,382 cm^{-1}, which can be attributed to the p_π-d_π transition in the W=O bond. The IR spectrum (in the 700–1000 cm^{-1} region) shows three absorption bands at 730, 800, and 955 cm^{-1}. The splitting of the 1080 cm^{-1} band (into 1040 and 1085 cm^{-1} bands) is attributed to the stretching of the P—O bond. Isotopic exchange takes place in the case of the tungsten atom, but not in the case of the phosphorus atom. An X-ray diffraction study shows that the U(IV) heteropolytungstate belongs to the hexagonal system with a = 9.018 Å and $\alpha = 80°41'$. The stability constant, determined by paper electrophoresis, is 2.22×10^6. Salts of other cations may be obtained by double exchange reactions in aqueous solutions. The acid form of $K_{16}[U(P_2W_{17}O_{61})_2]$ can be obtained in aqueous solution, and it remains stable for ten days.

C. POTASSIUM BIS(UNDECATUNGSTOBORATO)THORATE(IV)

$$Th(NO_3)_4 + 2\ H_3BO_3 + 22\ Na_2WO_4 + 26\ CH_3COOH \longrightarrow Na_{14}[Th(BW_{11}O_{39})_2] + 4\ NaNO_3 + 26\ CH_3OOH + 16\ H_2O$$

■ **Caution.** *Thorium is a radioactive element and must be manipulated with appropriate care. See caution notice in section A.*

Sodium tungstate dihydrate (25 g, 0.075 mole) is dissolved in 100 mL water and the pH is adjusted to 6.5 using glacial acetic acid. Then, 2.5 g (0.04 mole) of H_3BO_3 is added and the solution is heated for 30 minutes. A solution containing 2.1 g (0.0038 mole) of $Th(NO_3)_4 \cdot 4H_2O$ is added with continuous stirring, and the heating is continued for another 15 minutes. The solution is cooled to room temperature and filtered. Then, 10.2 g (0.136 mole) of finely ground KCl is added to the filtrate, whereupon a white, crystalline precipitate begins to deposit, which is allowed to stand at 5° overnight. The white precipitate is recrystallized for the minimum quantity of warm water (70°), and the final product is filtered, washed with small quantity of cool water, and dried in air.[17] The yield is about 37%. *Anal.* Calcd. for $K_{14}[Th(BW_{11}O_{39})_2] \cdot 38H_2O$: K, 8.08; Th, 3.42; W, 59.67; H_2O, 10.09. Found: K, 8.12; Th, 3.39; W, 60.04; H_2O, 9.86.

Properties

The UV absorption spectrum shows maxima at 39,572 and at 49,737 cm^{-1}. The IR absorption spectrum shows maxima at 760, 875, and 960 cm^{-1}. The absorption band at 1240 cm^{-1} is attributed to the B—O bond. X-ray diffraction shows that $K_{14}[Th(BW_{11}O_{39})_2] \cdot 38H_2O$ has a face-centered cubic structure, with a = 18.407 Å. Salts containing other cations can be obtained by double-exchange reactions in an aqueous medium.

D. POTASSIUM BIS(UNDECATUNGSTOPHOSPHATO)THORATE(IV)

$$Th(NO_3)_4 + 2\ K_7PW_{11}O_{39} \longrightarrow K_{10}[Th(PW_{11}O_{39})_2] + 4\ KNO_3$$

■ **Caution.** *Thorium is a radioactive element and must be manipulated with appropriate care. See caution notice in Section A.*

A solution containing 0.5 g (0.001 mole) of $Th(NO_3)_4 \cdot 4H_2O$ in 20 mL of water is added slowly and with continuous stirring to a warm solution (60°) of 5.7 g (0.002 mole) of $K_7[PW_{11}O_{39}] \cdot 14H_2O$ in 75 mL of water. The mixture is maintained at 60° with stirring for another 10 minutes, is then cooled to room temperature, filtered, and allowed to stand. After a few days, well-shaped, white crystals form, which are separated from the solution. These crystals are filtered, washed with a small quantity of cool water, and dried in air.

The yield is about 32%. The product may be recrystallized from the minimum quantity of warm water (60°) at pH 5.0–5.5. *Anal.* Calcd. for $K_{10}[Th(PW_{11}O_{39})_2] \cdot 36H_2O$: K, 5.90; Th, 3.50; P, 0.93; W, 61.04; H_2O, 9.78. Found: K, 5.88; Th, 3.45; P, 0.91; W, 60.70; H_2O, 9.89.

Properties

The UV absorption spectrum shows bands at 39,520 and 49,640 cm^{-1}. The IR spectrum shows absorption maxima at 752, 862, and 950 cm^{-1}. The splitting of the absorption band, which is attributed to the stretching of the P—O bond, occurs at 1042 and 1086 cm^{-1}. Salts containing other cations may be obtained by double exchange reactions in aqueous solution. The potassium salt of Th(IV) heteropolytungstate crystallizes in the cubic system.

E. POTASSIUM BIS(HEPTADECATUNGSTODIPHOSPHATO)THORATE(IV)

$$Th(NO_3)_4 + 2\,K_{10}[P_2W_{17}O_{61}] \longrightarrow K_{16}[Th(P_2W_{17}O_{61})_2] + 4\,KNO_3$$

■ **Caution.** *Thorium is a radioactive element and must be manipulated with appropriate care. See caution notice in Section A.*

A solution of 0.25 g (0.0005 mole) of $Th(NO_3)_4 \cdot 4H_2O$ in 20 mL of water is added in small quantities and with stirring to a solution of 4.8 g (0.001 mole) of $K_{10}[P_2W_{17}O_{61}] \cdot 23H_2O$ in 75 mL of warm water (60°); this temperature is maintained for another 15 minutes. The reaction mixture is cooled to room temperature, filtered, and allowed to stand for three to four days. Fine, white crystals are separated from the solution, filtered, washed with small quantities of cool water, and dried in air.

The yield is 47%. The product is recrystallized from the minimum quantity of warm water (60°) at pH 5.0–5.5. *Anal.* Calcd. for $K_{16}[Th(P_2W_{17}O_{61})_2] \cdot 45H_2O$: K, 6.25; Th, 2.32; P, 1.24; W, 62.54; H_2O, 8.10. Found: K, 6.42; Th, 2.35; P, 1.33; W, 62.80; H_2O, 8.19.

Properties

The UV absorption spectrum is similar to that of the unsaturated heteropolyanion $K_{10}[P_2W_{17}O_{61}]$, but it shows absorption maxima that are less distinct and that are shifted to lower energies (39,690 and 49,480 cm^{-1}). The IR spectrum shows absorption maxima at 750, 862, and 956 cm^{-1}. The splitting of the band, attributed to the P—O bond, occurs at 1038 and 1092 cm^{-1}. Salts containing other cations may be obtained by double-exchange reactions in aqueous medium. As with other compounds of this class, the Th(IV) heteropolytungstate has the Dawson structure.

References

1. L. C. W. Baker and T. P. McCutcheon, *J. Am. Chem. Soc.*, **78,** 4503 (1956).
2. L. C. W. Baker, V. Simmons Baker, K. Eriks, M. T. Pope, M. Shibata, O. W. Rollins, J. H. Fang, and L. L. Koh, *J. Am. Chem. Soc.*, **88,** 2329 (1966).
3. P. A. Souchay, *Ions Mineraux Condenses,* Masson, Paris, 1969.
4. T. J. R. Weakley and S. A. Malik, *J. Inorg. Nucl. Chem.*, **29,** 2935 (1967).
5. M. Tourné and G. Tourné, *Bull. Soc. Chim. France,* **4,** 1124 (1969).
6. R. D. Peacock and T. J. R. Weakley, *J. Chem. Soc. (A)*, 1836 (1971).
7. R. Contant and I. P. Ciabrini, *J. Chem. Res.*, (S)222, (1977); (M)2601 (1977).
8. V. I. Spitzyn, M. M. Orlova, O. P. Saprykina, A. S. Saprykin, and N. N. Krot, *Zh. Neorg. Khim.*, **22,** 2508 (1977).
9. R. K. C. Ho and W. G. Klemperer, *J. Am. Chem. Soc.*, **100,** 6772 (1978).
10. W. H. Knoth, *J. Am. Chem. Soc.*, **101,** 759 (1979).
11. W. H. Knoth, *J. Am. Chem. Soc.*, **101,** 2211 (1979).
12. A. Botar and T. J. R. Weakley, *Rev. Roumaine Chim.*, **18,** 1155 (1973).
13. Gh. Marcu, M. Rusu, and A. Botar, *Rev. Roumaine Chim.*, **19,** 827 (1974).
14. Gh. Marcu and M. Rusu, *Rev. Roumaine Chim.*, **21,** 385 (1976).
15. Gh. Marcu and I. Ciogolas, *Rev. Roumaine Chim.*, **24,** 1049 (1979).
16. Gh. Marcu and I. Duca, *Rev. Roumaine Chim.*, 1981 (in press).
17. J. H. Kennedy, *Anal. Chem.*, **32,** 150 (1960).

39. A GENERAL SYNTHESIS FOR GOLD(I) COMPLEXES

Submitted by A. K. AL-SA'ADY,* C. A. McAULIFFE,* R. V. PARISH,* and J. A. SANDBANK*
Checked by R. A. POTTS† and W. F. SCHNEIDER†

Large numbers of gold(I) complexes are reported in the literature, most involving ligands with the donor atoms phosphorus, arsenic, and sulfur. In preparing such

*Department of Chemistry, The University of Manchester Institute of Science and Technology, Manchester M60 1QD, United Kingdom.
†Department of Natural Sciences, University of Michigan—Dearborn, Dearborn, MI 48128.

complexes, it is usually necessary to start from gold(III), in the form of sodium tetrachloroaurate(III) or the corresponding acid, which then requires a reduction step. In most cases the ligand itself functions as the reductant, following the classical procedure of Levi Malvano,[1] causing one or two molar equivalents of the ligand molecules to be consumed in addition to those that are coordinated in the product (Equations 1–3).

$$2\ R_3P + Na[AuCl_4] + H_2O = [AuCl(PR_3)] + R_3PO + 2\ HCl + NaCl \quad (1)$$

$$2\ R_2S + H[AuCl_4] + H_2O = [AuCl(SR_2)] + R_2SO + 3\ HCl \quad (2)$$

$$3\ RSH + Na[AuCl_4] = [AuSR] + RSSR + 3\ HCl + NaCl \quad (3)$$

If the ligand is inexpensive and its oxidized form has suitable solubility, this method is quite convenient. However, some ligands are expensive or must be obtained by complicated syntheses, or the complex may be difficult to separate from the oxidized ligand. In such circumstances, other reducing agents may be used, such as sulfur dioxide[2] or sulfite,[3] but reduction is slow and difficult to control. The synthesis described here makes use of thiodiglycol (2,2′-thiodiethanol) as an inexpensive, odorless reductant that yields a stable aqueous solution containing chloro(thiodiglycol)gold(I), equation (2), where R = CH_2CH_2OH.[4] The desired ligand is then added, undiluted (if a liquid), or dissolved in water or alcohol, which usually precipitates the complex in quantitative yield (equations 4–5).

$$RSH + AuCl(\text{thiodiglycol}) = [AuSR] + \text{thiodiglycol} + HCl \quad (4)$$

$$R_3P + AuCl(\text{thiodiglycol}) = [AuCl(R_3P)] + \text{thiodiglycol} \quad (5)$$

This method is thus rapid, clean, and high-yielding, and may involve any ligand for which the complex has a lower solubility than the free ligand.

Procedure

Sodium tetrachloroaurate(III) dihydrate* (0.397 g, 1.0 mmole) is dissolved in water (10 mL), and the solution is *cooled in ice*. To this solution is added the thiodiglycol (0.366 g, 3.0 mmoles), undiluted, with stirring. This addition must be made very slowly (45 minutes) and may be stopped when the yellow color of the solution is discharged. A solution of the ligand, 4-ethylbenzenethiol (0.140 g, 1.0 mmole), in chloroform (20 mL) is added dropwise with stirring over 20 minutes. The chloroform layer is separated and added dropwise to methanol (40 mL), and the pale yellow complex, 4-ethylbenzenethiolatogold(I) (0.33 g, 98%), precipitates. The product is filtered, washed with methanol, and dried

*This salt is somewhat hygroscopic and should be weighed in a dry atmosphere.

under vacuum. *Anal.* Calcd. for C_8H_9AuS: C, 28.7; H, 2.7; S, 9.6; Au, 59.0. Found: C, 28.8; H, 2.7; S, 9.5; Au, 59.0.

If the product precipitates directly from the aqueous solution, the thiol may be used undiluted, but the complex must then be carefully dried. This method is appropriate for aliphatic thiols and for benzenethiol, the gold(I) derivatives of which are insoluble in most solvents.

Complexes of neutral ligands may be prepared similarly. To a gold(I) solution prepared as above is added a solution of the ligand $Ph_2P(CH_2)_2O(CH_2)_2O(CH_2)_2PPh_2$ (poop, 0.245 g. 1.0 mole) dissolved in ethanol (10 mL). An immediate white precipitate of the complex [ClAu(poop)AuCl] is formed (0.94 g, 98%). *Anal.* Calcd. for $C_{30}H_{32}Au_2Cl_2O_2P_2$: C, 37.9; H, 3.4; P, 6.5. Found: C, 37.9; H, 3.5; P, 6.7.

These complexes of neutral ligands can also be prepared by adding the ligand in chloroform solution, but, if the product is soluble in chloroform (as is usually the case with neutral organic ligands), recovery is often less than quantitative.

Preparation of Related Compounds

This procedure can be employed to prepare a wide range of complexes of the type AuSR, [AuCl(L)], and [ClAu(L′-L′)AuCl] with ligands such as: SR = SC_6H_4R' (R′ = 2-Me, 3-Me, 4-Me, 4-Et, 4-*i*-Pr, 4-*s*-Bu, 4-*t*-Bu, 2-COOH, 2-NH_2), $SCH_2C_6H_5$, SC_6H_{11}, $SC_{12}H_{25}$, $SC_{18}H_{37}$, *dl*-penicillamine, *l*-cysteine ethyl ester; L = Ph_3P, $(C_6H_{11})_3P$, Ph_3As; L′-L′ = $Ph_2P(CH_2)_nPPh_2$ (n = 2, 10), poop. The yield in all reactions is above 80% with the products being of good purity. The reaction with the ligand Ph_3Sb does not produce the gold(I) complex because the complex is known to be unstable[2] and the ligand is extremely insoluble in water.

Properties

The properties of these complexes depend to a great degree on the nature of the ligand. In general, the following are typical properties: colorless solid, powder, air stable, low solubility in most solvents, decomposition upon heating, characteristic Au-L stretching frequencing in the infrared.[2,5] Because of the expense of the starting material, everything containing gold should be collected and recycled by standard methods,[6] such as oxidation with aqua regia and then reduction to the metal with hydroquinone, sulfite,[3] or other suitable reducing agents.

References

1. M. Levi Malvano, *Atti Accad. naz. Lincei, Rc. Sed. Solen.*, **17,** 857 (1908).
2. C. A. McAuliffe, R. V. Parish, and P. D. Randall, *J. Chem. Soc. Dalton Trans.*, 1730 (1979).

3. S. Åkerstrom, *Arkiv Kemi,* **14,** 387 (1959).
4. G. Paret, *Ger. Pat.*, 1216296 (1965). A related method, using malodorous organosulfides, is described by H. M. Fitch, *Fr. Pat.*, 1367471 (1963).
5. G. E. Coates and C. Parkin, *J. Chem. Soc.*, 421 (1963).
6. C. F. Shaw and R. S. Tobias, *J. Chem. Educ.*, **49,** 286 (1972).

40. TRICHLORODIPHENYLANTIMONY(V)

Submitted by IONEL HAIDUC* and CRISTIAN SILVESTRU*
Checked by KURT J. IRGOLIC† and KENNETH A. FRENCH†

$$Sn(C_6H_5)_4 + SbCl_5 \longrightarrow SbCl_3(C_6H_5)_2 + SnCl_2(C_6H_5)_2$$

Trichlorodiphenylantimony(V) can be obtained from chlorodiphenylantimony-(III) and chlorine,[1] or sulfuryl chloride,[2] or by heating diphenylstibinic acid $(C_6H_5)_2SbO_2H$ with hydrochloric acid.[3,4] These reactions employ starting materials that are not readily available. Much more attractive are reactions of antimony pentachloride with phenylating reagents such as tetraphenyllead[5] and tetraphenyltin.[6] The simple reaction with tetraphenyltin gives good yields of pure diphenylantimony trichloride.

Procedure

Into a 1-L three-necked flask, equipped with a reflux condenser, a $CaCl_2$ drying tube, a dropping funnel, and a mechanical stirrer, are introduced 42.6 g (0.1 mole) of tetraphenyltin and 300 mL of carbon tetrachloride. A solution of 30.0 g (23.8 mL, 0.1 mole) of antimony pentachloride in 30 mL is dropped into the stirred, refluxing suspension of tetraphenyltin. After all the $SbCl_5$ has been added, the mixture is refluxed for two hours. During this time the solution becomes gray and contains a fine precipitate. After cooling to room temperature, the mixture is filtered to give 38 g of crystalline, impure $SbCl_3(C_6H_5)_2$ (mp 168–70°). The filtrate is concentrated on a rotary evaporator to a volume of 25–30 mL. After cooling, an additional 9 g of $SbCl_3(C_6H_5)_2$ is obtained. Dichlorodiphenyltin(IV) dichloride can be recovered from the filtrate.

Trichlorodiphenylantimony(V) is recrystallized from 650 mL of 5 *M* hydrochloric acid. The hot solution is filtered from a small insoluble residue through a Büchner funnel (9-cm, Whatman #1 filter) or a coarse glass-fritted funnel of

*Chemistry Departent, Babes-Bolyai University, R-3400 Cluj-Napoca, Roumania.
†Chemistry Department, Texas A & M University, College Station, TX 77843-1243.

similar size under an aspirator vacuum. On cooling, the filtrate deposits 31.6–35.4 g (79–88% yield) of fine, crystalline $SbCl_3(C_6H_5)_2 \cdot H_2O$ (mp 175°).

Properties

Trichlorodiphenylantimony(V) monohydrate forms needle-like crystals which melt at 175° (lit. 176°.[5] It is soluble in ethanol, hot hydrochloric acid, and acetone, moderately soluble in benzene, and slightly soluble in carbon tetrachloride.[5]

An X-ray structure determination[7] established octahedral coordination for antimony in $SbCl_3(C_6H_5)_2 \cdot H_2O$. The compound can be dehydrated by heating under vacuum. The anhydrous compound is a chlorine-bridged dimer as established by X-ray diffraction.[8]

When dissolved in methanol, the compound hydrolyzes partially to form $[(C_6H_5)_2SbCl_2]_2O$.[9] Treatment with aqueous alkaline solutions produces diphenylstibinic acid $(C_6H_5)_2SbO_2H$.[3]

References

1. A. Michaelis and A. Gunther, *Chem. Ber.*, **44,** 2316 (1911).
2. N. Nishii, Y. Matsumura, and R. Okawara, *J. Organomet. Chem.*, **30,** 59 (1971).
3. H. Schmidt, *Liebigs Ann. Chem.*, **429,** 123 (1922).
4. O. A. Reutov, *Doklady Akad. Nauk SSSR*, **87,** 991 (1952).
5. A. E. Goddard, J. N. Ashley, and R. B. Evans, *J. Chem. Soc.*, **121,** 978 (1922).
6. (a) C. Silvestru, Thesis, Babes-Bolyai University, Cluj-Napoca 1980; (b) S. N. Bhattacharyya, I. Husain, and P. Ray, *Indian J. Chem.*, **19 A,** 594 (1980).
7. T. H. Polynova and M. A. Porai-Koshits, *Zhur. Strukt. Khim.*, **7,** 642 (1966).
8. J. Bordner, C. C. Doak, and J. R. Peters, *J. Am. Chem. Soc.*, **96,** 6763 (1974).
9. L. Kolditz, M. Gitter, and E. Roesel, *Z. anorg. allg. Chem.*, **316,** 270 (1962).

41. TUNGSTEN AND MOLYBDENUM TETRACHLORIDE OXIDES

Submitted by A. J. NIELSON*
Checked by R. A. ANDERSEN†

$$MO_3 + SOCl_2 \longrightarrow MOCl_4$$
$$(M = W, Mo)$$

*Department of Chemistry, University of Auckland, Auckland, New Zealand.
†Department of Chemistry, University of California, Berkeley, CA 94720.

Tungsten tetrachloride oxide is a convenient starting material for the synthesis of a variety of tungsten complexes containing oxo[1] and imido[2]-ligands. It has been prepared by refluxing tungsten trioxide with octachlorocyclopentane,[3] by sealed tube reactions of tungsten hexachloride with WO_3[4] or dry sulfur dioxide,[5] and by the chlorination of WO_3.[6] A simple preparation involves refluxing WO_3 with sulfinyl chloride.[7] The reaction has also been carried out in a bomb.[8]

Molybdenum tetrachloride oxide has been used in the synthesis of oxo complexes of molybdenum(VI) and (V).[9] It has been prepared by reaction of $MoCl_5$ with MoO_2Cl_2[10] or MoO_3,[11] and by subliming it away from the solid material obtained by treating carbon tetrachloride solutions of $MoCl_5$ with oxygen.[7] The compound is also prepared by refluxing MoO_3,[12] sodium molybdate or MoO_2Cl_2 with sulfinyl chloride.[7]

For large-scale preparations of the tetrachloride oxides, the reaction of the metal trioxide with sulfinyl chloride is most convenient. The procedure outlined below is similar to that of Colten and co-workers,[7] but gives details for obtaining the compounds, particularly tungsten tetrachloride oxide, in excess of 50-g quantities. The reactions may be scaled up or down several fold without deleterious effect.

Materials and General Procedure

Molybdenum and tungsten trioxides are available from most chemical suppliers. If commercial samples are found to be relatively unreactive, MoO_3 may be freshly prepared from ammonium molybdate[13] by thermal decomposition at around 200°. WO_3 may be obtained by acidifying a solution of sodium tungstate with HCl, filtering the resultant precipitate, washing with water, and drying at 100°. Commercial sulfinyl chloride should be distilled prior to use; otherwise, brown-colored solutions result. All reactions should be carried out in an efficient fume hood, and all manipulations involving the solid oxytetrachlorides should be carried out under dry, oxygen-free nitrogen, using normal techniques for air-sensitive compounds.

Procedure

■ **Caution.** *Some of the substances in this synthesis (e.g., sulfinyl chloride) are volatile and toxic, and must be handled with care in an efficient fume hood.*

A. TUNGSTEN TETRACHLORIDE OXIDE

Finely ground WO_3 (100 g, 0.43 mole) is placed in a three-necked 2-L flask and 1 L of sulfinyl chloride is added. The flask is fitted with a stopcock, a stopper, and two spiral reflux condensers in tandem with all glass attachments (wired down to prevent forceful ejection). The mixture is refluxed at a rate such that violent bumping is avoided, until most of the WO_3 is consumed, or for at least two weeks, if commercial samples are somewhat unreactive. If bumping proves troublesome, a mechanical stirrer should be used. More sulfinyl chloride is added if the volume of solution decreases on prolonged reflux, and the condenser mouth is wiped periodically to prevent accumulating droplets from falling into the reaction mixture.

The heat is turned down and any unreacted WO_3 is allowed to sink to the bottom of the orange-red solution. While the solution is warm, the reflux condensers are removed, a rubber septum is inserted, and the flask is flushed with nitrogen. The liquid is transferred to another 2-L, two-necked flask (previously fitted with a stopcock, serum cap, and vent needle, and filled with nitrogen), by way of a stainless steel transfer tube. The transfer is carried out carefully, using a positive nitrogen pressure, until the solution begins to carry WO_3 with it. The receiving flask should be kept warm with a heating mantle, since cooling causes precipitation of $WOCl_4$. If WO_3 contaminates the warm solution, it should be allowed to settle, and the solution should be transferred to another flask. (Alternatively, the solution is filtered under N_2 while hot. Cold glass filtering devices will cause precipitation of $WOCl_4$ and clogging of the sinter.)

The solvent is removed under vacuum, leaving the compound as orange-red crystals. When dry, the solid is finely ground with a mortar and pestle in a glove bag, and placed in a Schlenk flask, which is then evacuated for several hours. The yield of tungsten oxytetrachloride is about 120–135 g, 80–90%. The solid may smell slightly of sulfinyl chloride, but it is of sufficient purity for most uses. A higher-purity material may be obtained by sublimation at 140° and 5–10 mm Hg. *Anal.* Calcd. for $WOCl_4$: Cl, 41.5; W, 53.8. Found: Cl, 41.2; W, 53.1.

B. MOLYBDENUM TETRACHLORIDE OXIDE

Molybdenum trioxide (30 g, 0.21 mole) is refluxed under nitrogen with 500 mL of sulfinyl chloride for 8–12 hours, or until no white solid remains. The cooled solution is filtered under nitrogen through a sintered-glass filter, and the solvent is removed under vacuum, to give the compound as a dark-green solid. The material is held under vacuum for several hours to remove vestiges of sulfinyl chloride, and is stored in a Schlenk flask under nitrogen at $-20°$. The yield is approximately 51 g, 97%. The compound is sufficiently pure for most uses. A higher-purity material may be obtained by sublimation at 50–60° and 10^{-3} torr. *Anal.* Calcd. for $MoOCl_4$: Cl, 55.9; Mo, 37.8. Found: Cl, 55.4; Mo, 37.3.

Properties

Tungsten tetrachloride oxide is an orange-red, air- and moisture-sensitive solid melting at 211°.[3] It is insoluble in aliphatic hydrocarbons, but dissolves appreciably in aromatic and chlorinated solvents, to give dark-red solutions. Its 1:1 adducts form in coordinating solvents, such as THF, MeCN, etc. The compound does not mull in Nujol, and it reacts with HBr and CsI. The crystal structure[14] shows that $WOCl_4$ is tetragonal, with $WOCl_4$ square pyramids weakly associated into chains through W· · · ·O interactions through the basal plane. The compound may be stored at room temperature under nitrogen for many months, but vacuum grease in glass joints should be renewed periodically.

Molybdenum tetrachloride oxide is a very moisture-sensitive, green solid, melting at 101–103° to give a brown liquid. A red-brown vapor is given off at about 120°.[12] The compound is soluble in aromatic and chlorinated solvents, forming adducts with coordinating solvents. It does not mull in Nujol, and reacts with HBr or CsI. In carbon tetrachloride, ν (MoO) occurs at 1009 cm^{-1}.[15] The crystal structure[16] determination shows an $MoOCl_4$ square pyramid, weakly associated with one other $MoOCl_4$, through Mo· · · · ·Cl interactions through the basal plane, suggesting incipient dimer formation. The $MoOCl_4$ is reduced photolytically and thermally to $MoOCl_3$ and chlorine,[15] and should be stored at $-20°$ in the dark. Long storage should be avoided, and fresh samples should be prepared prior to use.

References

1. (a) S. P. Anand, R. K. Multani, and B. D. Jain, *J. Organomet. Chem.*, **17,** 423 (1969); (b) H. Funk, W. Weiss, and G. Mohaupt, *Z. anorg. allg. Chem.*, **304,** 238 (1960); (c) A. V. Butcher, J. Chatt, G. J. Leigh, and P. L. Richards, *J. Chem. Soc. Dalton,* 1064 (1972); (d) S. P. Anand, R. K. Multani, and B. D. Jain, *Bull. Chem. Soc. Jap.*, **42,** 3459 (1969).
2. Imido ligands are not included in this synthesis herewith.
3. S. E. Feil, S. Y. Tyree, and F. N. Collier, *Inorg. Syn.*, **9,** 123 (1967).
4. J. Tillack, *Inorg. Syn.*, **14,** 109 (1973).
5. G. W. A. Fowles and J. L. Frost, *J. Chem. Soc.* (*A*), 671 (1967).
6. A. Michael and A. Murphy, *Am. Chem. J.*, **44,** 382 (1910).
7. R. Colton and I. B. Tomkins, *Aust. J. Chem.*, **18,** 447 (1965).
8. H. Hecht, G. Jander, and H. Schlapmann, *Z. anorg. allg. Chem.*, **254,** 255 (1947).
9. (a) N. K. Kaushik, R. P. Singh, H. S. Sangari, and G. S. Sodhi, *Syn. React. Inorg. Metal-Org. Chem.*, **10,** 617 (1980); (b) S. K. Anand, R. K. Multani, and B. D. Jain, *J. Ind. Chem. Soc.*, **45,** 1130 (1968).
10. W. Puttbach, *Liebig's. Ann.*, **201,** 125 (1880).
11. I. A. Glukov and S. S. Eliseev, *Zhur. Neorg. Khim.*, **7,** 81 (1962).
12. R. Colton, I. B. Tomkins, and P. W. Wilson, *Aust. J. Chem.*, **17,** 496 (1964).
13. J. Deluca, A. Wold, and L. H. Brixner, *Inorg. Syn.*, **11,** 1 (1968).
14. H. Hess and H. Hartung, *Z. anorg. allg. Chem.*, **344,** 157 (1966).

15. M. L. Larson and F. W. Moore, *Inorg. Chem.*, **5,** 801 (1966).
16. J. C. Taylor and A. B. Waugh, *J. Chem. Soc. Dalton,* 2006 (1980).

APPENDIX

SPECIAL HAZARD NOTICE

PREPARATION OF TETRAMETHYLDIPHOSPHINE DISULFIDE AND ETHYLENEBIS(DIMETHYLPHOSPHINE) (DMPE)

J. E. Bercaw* and G. W. Parshall†

Two serious accidents[1,2] have occurred during the synthesis of tetramethyldiphospine disulfide described in *Inorganic Syntheses,* Vol. 15, pp. 186–187 (1974). In both instances a rapid temperature rise overpressurized the reaction flask, which exploded and sprayed the laboratory with glass fragments. While the exact cause of the accidents has not been established, it appears likely that a high concentration of insoluble products can lead to difficulties in temperature control if stirring is inadequate.

It is recommended that the *Inorganic Syntheses* procedure be abandoned in favor of the older *Organic Syntheses* procedure[3] for the synthesis of tetramethyldiphosphine disulfide. Although the procedures are similar, the *Organic Syntheses* procedure is carried out in a more dilute solution and no problems have been reported. The following precautions are strongly urged:

1. The phosphorus trichloride sulfide ($PSCl_3$) should be distilled before use.
2. The reaction vessel should be cooled in an ice/salt bath (rather than in a dry ice/acetone bath) during the addition of the $PSCl_3$ solution to the Grignard reagent. The reaction temperature should be monitored carefully. If it falls below $-5°$, the addition should be stopped and the reaction mixture cautiously rewarmed to 0–5° before resumption of the addition.
3. The reaction apparatus should be shielded throughout the addition of the $PSCl_3$ solution and the subsequent warming of the reaction mixture.

*Department of Chemistry, California Institute of Technology, Pasadena, CA 91125.
†Central Research and Development Department, E. I. du Pont de Nemours and Company, Inc., Wilmington, DE 19898.

If the ultimate goal of the synthesis is ethylenebis(dimethylphosphine) (DMPE), an alternative synthesis[4] based on ethylenebis(dichlorophosphine)[5] is recommended. This alternative procedure is convenient and is being adapted for publication in *Inorganic Syntheses*.

References

1. J. E. Bercaw, *Chem. Eng. News,* 30 April 1984, p. 4.
2. S. G. Davies, *Chem. Britain,* **20,** 403 (1984).
3. G. W. Parshall, *Org. Syn.,* **45,** 102 (1965); **Coll. Vol. 5,** 1016 (1973).
4. R. J. Burt, J. Chatt, W. Hussain, and G. J. Leigh, *J. Organomet. Chem.,* **182,** 203 (1979).
5. R. A. Henderson, W. Hussain, G. J. Leigh, and F. B. Normanton, *Inorg. Syn.,* **23,** 141 (1985).

INDEX OF CONTRIBUTORS

SUBJECT INDEX

Names used in this Subject Index for Volumes 21–25 are based upon IUPAC *Nomenclature of Inorganic Chemistry,* Second Edition (1970), Butterworths, London; IUPAC *Nomenclature of Organic Chemistry,* Sections A, B, C, D, E, F, and H (1979), Pergamon Press, Oxford, U.K.; and the Chemical Abstracts Service *Chemical Substance Name Selection Manual* (1978), Columbus, Ohio. For compounds whose nomenclature is not adequately treated in the above references, American Chemical Society journal editorial practices are followed as applicable.

Inverted forms of the chemical names (parent index headings) are used for most entries in the alphabetically ordered index. Organic names are listed at the "parent" based on Rule C-10, *Nomenclature of Organic Chemistry,* 1979 Edition. Coordination compounds, salts, and ions are listed once at each metal or central atom "parent" index heading. Simple salts and binary compounds are entered in the usual uninverted way, e.g., *Sulfur oxide* (S_8O), *Uranium(IV) chloride* (UCl_4).

All ligands receive a separate subject entry, e.g., *2,4-Pentanedione,* iron complex. The headings *Ammines, Carbonyl complexes, Hydride complexes,* and *Nitrosyl complexes* are used for the NH_3, CO, H, and NO ligands.

FORMULA INDEX

The Formula Index, as well as the Subject Index, is a cumulative index for Volumes 21–25. The Index is organized to allow the most efficient location of specific compounds and groups of compounds related by central metal ion or ligand grouping.

The formulas entered in the Formula Index are for the total composition of the entered compound, e.g., F_6NaU for sodium hexafluorouranate(V). The formulas consist solely of atomic symbols (abbreviations for atomic groupings are not used) and arranged in alphabetical order with carbon and hydrogen always given last, e.g., $Br_3CoN_4C_4H_{16}$. To enhance the utility of the Formula Index all formulas are permuted on the symbols for all metal atoms e.g., $FeO_{13}Ru_3C_{13}H_{13}$ is also listed at $Ru_3FeO_{13}C_{13}H_{13}$. Ligand groupings are also listed separately in the same order, e.g., $N_2C_2H_8$, 1,2-Ethanediamine, cobalt complexes. Thus individual compounds are found at their total formula in the alphabetical listing, compounds of any metal may be scanned at the alphabetical position of the metal symbol, and compounds of a specific ligand are listed at the formula of the ligand, e.g., NC for Cyano complexes.

Water of hydration when so identified is not added into the formulas of the reported compounds, e.g., $Cl_{0.30}N_4PtRb_2C_4 \cdot 3H_2O$.